D0120731

RFID
100 Success Secrets

RFID 100 Success Secrets - 100 Most Asked
Questions: The Missing Radio Frequency Identi-
fication Tag, Implementation and Technology
Guide

Rick Thomas

RFID 100 Success Secrets

Copyright © 2008

RFID 100 Success Secrets

There has never been an RFID Guide like this.

100 Success Secrets is *not* about the ins and outs of RFID. Instead, it answers the top 100 questions that we are asked and those we come across in forums, our consultancy and education programs. It tells you exactly how to deal with those questions, with tips that have never before been offered in print.

This book is also *not* about RFID's best practice and standards details. Instead, it introduces everything you want to know to be successful with RFID.

Table of Contents

The Edge and Downsides of Active RFID

Radio-frequency identification or RFID is the general term used in describing the new automatic identification technology which relies on the stored data and retrieving this information whenever the need arises. The RFID technology uses a device called the RFID tags. These tags are small objects that are applied or put on to any product, person, or animal to provide them with their unique identification through radio waves. The RFID tags can be grouped into three general varieties. These are the semi-passive, passive, and active. Among the three, the active RFID is the most unique.

Compared to the two other types of RFID tags, the active RFID has its own internal power source. This power is used to energize the integrated circuits as well as broadcasting the signal of the chip to the reader. The passive RFID, on the other hand, needs a battery in order to broadcast a signal and be read while the semi-passive RFID also has its own power source but it does not automatically broadcasts a signal.

Unlike the passive tags, the active RFID is more reliable and works with minimal errors. This RFID tag also has the ability to transmit messages at a higher level of powers making them more reliable in some areas that are hard for tracking. Active RFID is ideal when used under the water, humans, animals, metal, and long-distanced paces. This is also capable of responding strongly even in a weak power request.

The only downside of the active RFID is it bigger size and it's more expensive cost. Their potential shelf life is also shorter urging its manufacturers to produce only limited number of its type. However, the internal power source of active RFID is very reliable since it lasts up to 10 years.

Why Is An Active RFID Tag More Effective?

RFID means radio frequency identification. This technology uses transponders or RFID tags in order to store data which is then retrieved from a remote site. There are many benefits that enterprises see with the use of RFID and so they implement the use of this technology.

RFID tags are attached to the objects or the person the company or the agency is attempting to track. It may be inventory in warehouses, personnel in a security sensitive facilities or important and expensive equipment or asset of a company.

There are 3 varieties of RFID tags. The first are the passive RFID tags. This type has no power supply on its own so it only depends on the power given off by the antenna when it needs a response from the tags. It is therefore, only activated by the reader when the two are in proximity to the other. The semi-passive RFID tags, on the other hand, have their own batteries but their power supply does not allow the tags to send out signals. It only powers the microchip.

The third type is the active RFID tags. The major difference this type has with the passive RFID tags is that these tags have its own power supply. And they also differ with the semi-passive RFID tags in so far as these tags can power not only its integrated circuit by it can also transmit a signal to the reader. And because this type has its own power supply it can actually be more effective than the passive type. It can transmit higher levels of power and so an active RFID tag is able to transmit signals even in difficult situations such as that when in metal, in water and in farther distances.

Automotive RFID Solutions: Fuelling the Automotive Market into Hyper-drive

The global automotive RFID market is expected to post an annual growth rate of 20% as predicted by top market analysts. In actual monetary value, this could mean that the automotive RFID industries could generate hundreds of millions in net incomes. This has been fuelled by the need of car manufacturers for sophisticated identification solutions for their products and manufacturing processes. And RFID solutions have become the perfect choice of automotive giants because of the increasing sophistication of RFID products.

Another positive development in the automotive RFID market is the decreasing cost of RFID products. That is why more and more car manufacturers are adopting RFID technology for their businesses and production efforts. And costs of RFID solutions are expected to still go down as competition intensifies and new micro chip and miniaturization technology rolls out of the market. Automotive companies therefore would find it more economical to apply RFID for their production and manufacturing activities. Thus a continued pattern of growth could be expected in the long term as more companies migrate to RFID solutions.

Specifically, automotive companies could benefit from RFID solution by enhancing their corporate abilities to support supply and sourcing chains. RFID could also improve tracking of the production process from plant manufacturing and warehousing to product deliveries. Automotive RFID solutions could also increase corporate efficiency by automating data management, storage, retrieval, and monitoring. The significant reduction of manual bar coding could speed up production and greatly minimize errors and production wastage. All these benefits can only mean one thing for the automotive industry: more profit and further growth.

Why RFID Trumps Bar code

The usage of radio frequency identification is becoming more and more utilized nowadays compared to bar codes. In fact, many a research team reports that radio frequency identification tags also have enough of a capacity to actually store a lot of data to include the serial numbers, the individual product information as well as other types of information which different manufacturers would like to insert.

Unlike using the bar code system, radio frequency identification is able to scrutinize each and every product in an individual manner and does not really operate to identify the entire type of product. Radio frequency identification technology is a lot better than bar code system because it is easier to spot items or products within a given distance. Thus, one is not required to actually see the physical existence of such a product. These are plastic made and are passive tags which are actually a lot durable as well as reusable, thus contributing to the whole economizing feature of the technology.

Compared to bar code systems, these radio frequency identification transponders can also be embedded into your car tires for tire tracking, and these radio frequency identification chips may also be quite useful when it comes to cards that churn out electronic cash. It has even been said that such radio frequency identification tags may also be used to replace UPC or even the EAN bar codes because it has a lot of really good advantages which truly trump the erstwhile efficiency of the bar code system.

The Computer RFID Makes a Buzz

The world of technology will never be the same due to the launch of computer RFID. It is set to be one of a kind and totally different from the usual computers we may have been used to. The computer RFID is made literally as well as technologically tough. It can withstand the harshest weather conditions and you can even count on it to work even when submerged on water. It might be too good to be true but yes it is. The whole new innovation which involves computer RFID is not just some fantasy for some.

Aside from its durability, this new computer is also quite flexible when it comes to connecting wirelessly with other portals. You can expect it to be able to connect with various programs such as Bluetooth and Cisco. Because of its wireless capabilities, you can finally connect to various networks without the aid of going through a web of connections identified through octopus cable connections. Because it does not simply reply on cables and wires to transmit as well as receive data, the computer RFID can handle multiple applications hosting all at the same time. Thus the entire process becomes very efficient in terms of downgrading costs and also presents faster data transfers.

Speaking of downgrading, the said computer is also a certified space saver because of its size. It can easily fit into a certain place and it also has enhanced screen mechanisms to promote a sharper visual coverage. Most of all, it also has a touch screen display so users no longer have to rely on keyboard alone.

Importance of RFID to Consumers

Individuals get annoyed when people try to seek into their privacy. Some viewed RFID as an invasion of privacy, as the technology tries to identify the person and its whereabouts. It is not a surprised that some would not favour RFID on individuals, but RFID tags on products have become acceptable to consumers. With the RFID tag on products, consumers are assured of the product that they purchased. They are assured of the quality as having passed the company's standard of test and inspection. The RFID guarantees consumers that they are getting the best quality. It is for this reason that companies provide tags/labels to identify those that have gone through the process. RFID is everywhere. RFID technology can mostly be seen on bookstores and CDstores. The technology is an aid against anti-theft. You may notice that stores have RFID technology in place. The RFID tag attached to a product provides information as to whether the product has been tested, and sold. An alarm would sound if an item is taken out of the store if not considered as sold. It is the responsibility of the company to inform consumers that RFID technology is being used. Informing consumers and providing with notice is a best practice. This is to avoid misunderstanding between you and your customer. If for instance, a consumer opted to remove the RFID tag it should be clear to the consumer that such item loses its ability to return an item or benefit from a warranty or protection from law. With the Consumer RFID, consumers are protected and guaranteed excellent service.

Using the Counterfeit RFID for Better Society

During the later period of the last millennium, counterfeiting as a crime has flourished. This process is done through imitating a single item with an intention to deceive others that that faked item is the original one. This is not only true to currency, money, or documents. There were also counterfeiting in clothing, pharmaceuticals, software, watches, motorcycles, and cars. The end result of this deception includes danger to public and most of the times unsafe use of these products. Risk of life and property are highly involved with counterfeiting. But today, this devious deception can be prevented through the use of radio-frequency identification or RFID. Thus, the operation is called counterfeit RFID.

The use of RFID was proven to be effective in preventing counterfeiting. Since the early parts of this new century, businesses have lost millions of dollar due to counterfeiting. For example, counterfeited famous watch brands like Rolex are sold with much lesser price than the original one, attracting more and more buyers. Through the counterfeit RFID, the manufacturers can put on tags which will provide unique identification for every watch they have produced. Through that, consumers will know that the brand they have bought is the real brand and not the counterfeited version.

This benefit is not only for the jewelleries and clothing. Even medicines can use counterfeit RFID in order to detect what's the original from the counterfeited one. Through this, complications due to unsafe use of fake medicines can be prevented. Posting danger to human health will be lessened and consumers can now be assured that the drugs they take are really effective as they expect them to be.

RFID Implemented on Credit Cards

RFID technology has been making bigger and bigger innovations today. In fact, the RFID software has already hit the credit card market. MasterCard and Visa card holders will be able to use latest in RFID technology. The "Blink" cards have been developed for these credit card holders for use in their transactions. The idea behind these "Blink" credit cards is that a simple flick of the card will complete the transaction. The cards will be equipped with the RFID technology, allowing this easy feature for credit card holders. Millions of these "Blink" cards are being developed for Visa and MasterCard.

These "Blink" cards can be used in any service store which provides customers with RFID terminals to wave their cards at. Paying for something has never been easier, and faster, not to forget more fun. However, despite the good pros of this new technology of RFID, it still has, like many other inventions, a few cons to consider. The RFID terminals like more sensors have a minimum range requirement for it to be able to read a "Blink" card. Of course, this is actually a security measure as identity thefts would have an even harder time of stealing a card holder's information.

It is also said that the "Blink" cards will only be available for credit cards, and will not be made available for debit and bank cards. But that is not so much of a con as it is also considered that once a theft does get into your information, they would only be getting information on your credit card, and not your main bank account information. This technology is still fresh though, so further improvements are to be expected in the near future.

The Use of Electronic RFID in the Supply Chain Industry

The use of RFID since its introduction has branched out into different fields today. In the manufacturing and trade industry, the supply chain industry have benefited from the use of RFID. There is the introduction of the Electronic Product RFID or more formally known as Electronic Product Code Information Service, with the acronym EPCIS. This system allows supply chain partners and stores to track the products through the use of real time data. This is made possible by assigning unique serial numbers for each product. This way, both the supply chain and the store can have real time information, regardless of the time they used the same system. The EPC is a created serial that complements a designated bar code, unique to a product for proper identification.

With the introduction of EPC, EPC II was introduced next to comply with global standards and to meet the globalization trend that is prevalent in today's economy. In contrast to the EPC, it has an added feature which is the "dense of reader mode of operation". This prevents the interference of operations when 2 users are in proximity to the other. The EPC 2nd generation is very important because it works anywhere, globally and many manufacturers are making it. This means more competition among manufacturers which increases the volume; thereby the prices will also go down. This is an added advantage to those who want to buy this system.

EPC helps companies in a way that it provides a good tracking system for the "authentic" goods that they produce, in contrast to the "counterfeits" or "fake" goods. EPC also provides data that is very useful in the supply chain by providing data needed by the store and the supplier.

Fashion RFID Getting to Be a Norm Today

RFID is not just in use in the transportation sector, in the banking sector or in the government sector. Many businesses in their aim to achieve efficiency and productivity employ the RFID technology. Many enterprises have found its benefits to be invaluable. In warehouses, not only were they able to do tasks quite easily but they were also able to do it in a shorter amount of time. As many business people know, time is money; and so it that important to them. Whatever time they save means a better bottom line later.

But the use of RFID in warehouses and tracking is not limited to commodity products only. Even inventory of fashionable clothing has been known to take advantage of the technology. The importance of timely delivery of their products from the design table to the racks of the boutiques is quite vital to the success of their business. The sooner they have new lines on display the better chances of it selling well and in volume. For especially made to order dresses, time is of the essence. These clothes are usually ordered for specific occasion so it should arrive its destination before that day comes. This is made possible by the RFID technology. With it in place, tracking of the orders is made more efficient and making sure that orders go to the right persons is easier. Should one order be misplaced, tracking that specific dress is easier since there would be a record of where it would be at that exact moment. Should it still be in its original location then it can be found through the use of a portable reader. But still the time spent looking for it would be shorter. Because of this fashion RFID is becoming widely in place.

RFID technology: a sneak look on federal government's off beam insights

Since its development, RFID (radio frequency identification) technology was distinguished as a promising technology, particularly among large enterprise or companies. And true enough, its adoption by large businesses such as Wal-mart had resulted to an enhanced supply chain management—essential increased efficiency, increased cost-savings, and reduced possible errors, tampering, and theft.

RFID technology systems normally involve the integration of RFID tags to every product or commodity. An RFID tag stores essential data. These data are captured by a reader, and then transmitted into a designated machine or equipment. An RFID tag is similar to a bar code, however is distinct since it contains a microchip that only responds through radio wave query.

Apparently, RFID's capabilities as an advance technology solutions resulted for its increasing adoption by companies of various sizes. It has gained extensive popularity that even the federal government was not an exemption to recognizing its full potential. Some politicians even have strongly advocated for the federal government to get actively involved in further developing RFID technology. However, such advocacy among politicians failed to gain support from the community, and the technology and business sectors, but have instead drawn scepticism and apprehensions—reactions that are absolutely logical.

Indeed, for politicians to even consider being keenly involved in the design and implementation of RFID is nothing more than a futile foresight and effort. They should leave RFID matters to the experts. Otherwise, a propitious technology as such RFID would only end up as political propaganda; resulting for decisions and actions dictated by political pressures and preferences, instead of the marketplace and its primary concerned segments. And frankly speaking, many would perceive that nothing good can come out from such scenario.

File Tracking RFID Makes File Inventory Organized

Information is very important to businesses and the timely access and availability of information can be very vital to an enterprise. That is why companies would employ personnel and facilities just so they can collate and manage files of information very efficiently.

With RFID, files are can now easily be tracked. Companies should be able to organize them and see to it that everything is in its place. If the company implements RFID, they can automate their circulation of files within the company. And there's added security since RFID enables document control. The coming in and out of files is also regulated. This is very helpful since it's possible that files being checked in and out are more than just one. Without the help of this technology, the task of taking note of each transaction will be very tedious and time consuming. Because of this maintenance of the file inventory would be easier and faster. Fewer personnel would be needed and file inventory would be run more efficiently.

What's even supposed to be trickier is when a file is lost. But with an RFID tag on the file and with the use of a portable RFID reader, the search for the lost file will be easier and the chances of finding the file are better. If the file has been filed incorrectly it would be detected by the reader and the mistake will be easily corrected. Since the personnel doing this task wouldn't have to go through the files one by one, it will also save them time.

All About Fresh Patents RFID

Lots of ventures have already seen the usability of RFID. Because of these, both private and government owned companies have started to utilize the said device according to their needs. Nowadays, you can pretty much see lots of different uses for RFID and there are various industries which actually utilize the said program. Because of this, fresh patents RFID have emerged out of the growing IT need.

As more and more business sectors realize the potential that RFID can bring in for them, more and more software developers have become attracted to turn this chip in many variations. As such, users and potential customers can find themselves in hodge-podge of choices without the proper corroborating means. This is where fresh patents RFID comes in. It has its home base on the internet and it describes in detail the many different versions which are now available in the field of RFID technological development.

Because of fresh patents, users can now see the latest innovations regarding the said chip. They can compare prices with other products and choose the ones which closely fit their business needs. There are also product reviews involved which can help the user make up his or her mind before purchasing the said chip. In more ways than one, the fresh patents RFID helps the potential clientele benchmark according to what they truly need for their business to prosper. Amidst the overwhelming amount of choices, they can still seek the ones which they really need and which they will be comfortable in using.

The Impacts of RFID in the Global Business

More usual that not, the growth in the economy in any country requires an equally developed technology in order to sustain its unprecedented progress. Clearly visible, developed countries like most of the European parts, the United States, and the giant country of the Asia-Pacific, Japan, make full utility of its advanced technology not only to sustain its development but also to extend assistance to its business sector to make sure that the whole country and its people continue to enjoy the luxuries of life and inevitably maintain its rule in the whole region. To these countries, technology means competitiveness in the global business and economy.

One of the most common forms of latest technology being used by most developed countries is the radio-frequency identification or RFID. It is a method that utilizes RFID tags or commonly termed as transponders to clearly identify almost everything – animals, people, materials, goods, and all sorts that can be identified. The use of the radio waves makes the whole idea of RFID possible. This mechanism allows for a more safer, secure, and expeditious way of tracking and monitoring things, activities, and even people that are employed in any business.

You may have not noticed it, but generally of the things that we are actually doing in our environment requires the use of the RFID. This is what the global business requires everyone to deal with. For example, when you secure your passport for travel purpose, the passport issued to you has employed the RFID technology. Even when you go to bigger schools and universities, the use of the RFID method is widely evident or even when you transact with your pharmaceutical people.

The impact of RFID is very huge and everyone must respond to it. Otherwise, you shall never be able to suit yourself in the global perspective.

Infant RFID: Little Angels Safe from Abductions

The provider of safety and identification technology, VeriChip Corporation, has employed the technological radio frequency identification (RFID), to avoid the abduction of infants. There are actually 233 cases of infant abductions that occurred in US for the last 22 years. Half of these cases happened inside the health care facilities.

VeriChip's "Hug" is essentially an infant RFID system that protects the infants by sounding a loud alarm and by flashing a warning on the monitor in the station of the nurses. When the parents of the baby tried to get their baby without asking permission from the nursery of the hospital, the alarm will automatically turn on, indicating a warning to the hospital staffs.

There is actually the so-called "Code Pink" alert, which will deliver the warning to the security officials who can stop the abduction. Hence, the infant can be recovered unharmed and be returned to the staff of the maternity ward safely. This technology, "hugs" or infant RFID system, was made possible through the use of monitoring software including the ankle bracelet that comprises a small radio transmitter. This transmitter was purposely designed to keep the infants safe from being detached from the facility of health care without permission.

There are numerous hospitals nowadays that give the infants Hugs tag in the form of ankle or wrist bracelet to observe the movement within the hospital. The exit ways in these hospitals too are equipped with electrical monitor to sense unauthorized removal of the baby.

Additionally, this infant RFID also avoid the incident of the infant mismatching, which can sometimes happen accidentally. These days, there are about 900 hospitals here in US that use this technology.

RFID Information Technology Process

Companies today have become aware of the existence of the RFID technology and what good it does in terms of its ability to identify products, track products whereabouts, increased productivity of workforce and savings that it brings. Because of this, many service companies integrator are available in the market that caters to have the RFID implementation work. These service companies seek to integrate the issues affecting the business and try to find solutions using the RFID technology. The following are the RFID information technology process that service integrators provide: 1.   They   assess   the   business current procedures, environmental conditions, and business requirements. The client's facility is surveyed to determine the potential physical barriers and radio frequency trouble spots. 2. They design or customized a RFID solution that is based on the business requirements and at the same time necessary to meet their client's needs. They evaluate and select the best tag option to apply. They determine the right RFID label and the best location to place such label to achieve optimal read performance. 3. They implement the installation of hardware systems and integrating software within the site location. The process starts with a pilot program and eventually rolls out to the rest of the organization. 4.   They   provide   continuous service of equipments, media replacement, and maintenance of product's lifecycle 5.   They provide the training for the client's staff and assess the system continuously if refinement is necessary. You would certainly need help from an RFID Information Technology before you can start on adapting RFID technology in your business. RFID technology may be costly at start, but its rate of investment out grows the expenses incurred.

Jobs RFID Technology Will Grow with Implementation

The use of RFID technology applications has helped many enterprises and agencies improve the quality of their work. For businesses, this means more productivity and of course, additional income. For some, it could be added security. Work is done in a shorter span of time and control is put in place. This is because RFID technology allows identification to be automated and to be done without any line of sight.

With these benefits known, more companies are encouraged to try putting RFID technology in place in their establishments. Hence, the demand for RFID has grown. And as long as and more companies are attempting to implement, there would also be a demand for RFID personnel. Jobs for RFID technology specialists would abound. And with many companies implementing the said technology, many companies would also need RFID people to run their systems and to maintain them from time to time. And since there quite so many applications for RFID, the need for professionals knowledgeable in it would also go up.

What's even attractive about this field is that there aren't many people specializing in it yet. With a high demand and a low supply, salaries would tend to be higher. And as the trend suggests, there would still be more applications and uses that will be developed in the future. This spells a good tomorrow for the RFID professionals. With such a need, they would not run out of jobs for a long time. Quite the contrary, more and more and companies would need them to update their systems instead.

Kidwear RFID Makes Detection for Lost Kids a Cinch

The term RFID stands for Radio Frequency Identification. This technology works as a tag, which may soon alter barcodes as a way of labelling the entire sorts of things. One of which is the clothing that every person can wear, more particularly for the kids. Through the use of this technology, it will be easier on the part of the parents to find the things that are needed by their kids. This RFID here is employed on kidswear.

Recently, there was a technology that was launched by Lauren Scott in California wherein the RFID was employed on the pyjamas of the kids. This kidwear rfid can truly be employed on different types of clothing. The very purpose of this technology is to guide the parents in locating their children if in case they were lost. This was made possible by the database that was established in the clothing of the kids. This database stores the essential information that concerns the kids. And once these kids get astray or are missing, all of the stored information will be delivered automatically to the law enforcement. Hence, locating the child is just a cinch.

The technological advancement like RFID is undoubtedly a good thing. Such technological advancement can help people in managing their accounts in easier way. Keeping the stock for others can also be effortless for every individual who has this. However, like the rest of the technology developments, this comprises a mixed up blessings as this can pose troubling risks to people's health, security, and privacy.

Boosting Business Sales through Labels RFID

The label plays a significant part for a certain brand to stand out. That is the primary reason why every businessman needs to give out the best to achieve a label that can generate a boost on the sales of their product. Essentially, the quality of label produces a great deal that concerns the product including the company it represents.

There is actually nothing to worry if the issue is the lack of creativity of the businessman himself to design the label. There are numerous labelling companies nowadays that can manage this work effectively. Many of these labelling companies make use of the most up to date flexographic presses that manufactures high-class labels that are also pressure-sensitive. More significantly, many of them are employing labels RFID. The latter stands for the radio frequency identification (RFID) smart labels, which is commonly used during the conversion process. The RFID media is commonly employed to achieve the best result and lessen greater expenses as well.

Finding the company that employs innovative way in using the RFID inlays are advantageous. This is because the selection of the finest RFID inlay is somewhat critical. There are still numerous components that are needed to be completed during the process to achieve the general success of RFID smart label course. This is the reason why one needs to be careful in choosing for the labelling company that will carry out his business label. The ideal label company is the one that ensure to deliver top performance for the planned application. Of course, the one that secures premier quality outcome.

Pushing for Legislation RFID

Most businesses even the government itself utilizes RFID for personal identification purposes. However, some people tend to be sceptical and quite apprehensive when it comes to using the said technology in affirming their personal data. Although the whole process proves to be more efficient and timely as compared to its old counterpart (the stamping days as well as the times when punches are tabbed onto forms), some people would still prefer the old rituals over the ease of RFID. And this is not without proper reason.

Some people and even professionals themselves in the field of IT are quite curious as to the receiving portals of RFID. Some say that the government may be storing these into some secret database or even selling them to big-shot secret agencies such as the FBI and even Interpol. Although these things remain to be a speculation, it cannot be denied that it could be a possible. Even if the government itself does not wilfully sell or submit these data into outside agencies, it is still fairly easy to intercept the said data.

So to help counter for this, legislation RFID has been made to quell the scepticism and make the public more attuned in receiving this new technological concept. Numerous bills have already been passed in order to make sure that these public becomes well taken care of however, the public is yet to see one solid legislation RFID which embodies what they want in terms of safety and security of personal identity. Perhaps, this is also because the technology remains to be at the mid-backseat of the market's consciousness.

Benefits of RFID in the Manufacturing Industry

Applications of the RFID technology in manufacturing firms have brought major benefits in both the supply chain management and factory operations. This has resulted to an increase in productivity, reduction of costs that allows a tracking of inventory, the reuse of containers, work in process and finished products.

The RFID in manufacturing has enabled the track and record the in-process assembly information using the RFID tag. An example is when an additional component is added to personal computer, this addition is recorded on the tag. At this point, the tag keeps the current inventory of the PC contents. This information would be read later and use for shipping list and invoice. The tag will also provide an information for personnel to take note during the installation and maintenance. The technology is ideal for those manufacturing firms with several product lines or with manufacture complex or customized products.

With the manufacturing RFID, personnel from the assembly line personnel could use an RFID reader to verify from which processes have been completed. They are able to determine which products are for inspections or tests. The technology enables them to automatically update the central production database as well as customer database.

Using the RFID in Manufacturing provides firms with real time tracking of inventory, part kits and sub-assemblies. The RFID tag will help personnel determine the product features, date manufacture, revision levels etc. It definitely reduces your cost, as you are not prone to errors due to the elimination of manual data entry and increases the productivity level of your personnel. Having RFID manufacturing would allow you to have more time to increase your production and meet the demands of your market.

Some Challenges in Marketing RFID Technology

Effective marketing of the RFID systems establishes branding and product recall for better market positioning. This could widen the consumer base of RFID products which could boost sales and push further growth of RFID companies. Top RFID companies are more effective in their marketing strategies outperforming start-ups and middle level manufacturers. However, some studies show that consumer appreciation of RFID systems is still low and more efforts will be needed to raise the awareness of the market.

This low level of RFID awareness shows that current brand leaders continue to struggle in terms of establishing the technology as a better data management alternative. Aggressively marketing RFID systems is seen as the key in overcoming some market misconceptions on the technology and raising its acceptance level among major industry players. Specifically, issues on privacy and security are the foremost problems seen by the market in any RFID deployment. Because the technology is relatively new, RFID data management and monitoring approach will certainly receive many wrong perceptions. It is therefore a challenge for the RFID industry especially brand and thought leaders to overturn negative market perceptions on the new technologies.

However, not everything is bleak in RFID market perception. The awareness and patronage of the system is growing in larger industries. This may be due to the fact that these industries need a cost effective and efficient way of managing their data solutions through economies of scale and superior performance. More large industries are deploying RFID technologies especially in their supply chain management and monitoring.

RFID: Improving Medical Accuracy

RFID has been used extensively in many fields especially in the supply chain and transportation. It helped a lot in the identification of many products and has also served as system of "passporting" for many establishments like in companies and some highly secured establishments like in the case of top government agencies.

But did you know that RFID also plays an important role in the medical field? Yes, RFID not only help increase medical accuracy, but it also helps in the diagnosis which in turn can help in the curing sickness. A lot of hospitals are now using barcode system in order to ensure that the patients are receiving the right treatment. With the use of information technology, the factor of human error is decreased significantly.

In the hospital setting, the use of RFID does not limit to tracking hospital assets such as beds, wheelchairs, incubators, and so on. RFID are now also used as identification for patients. Imagine in a nursery in an O.B. Ward wherein each baby has an identification badge which identifies each of them on whose child they are. This decreases the chance that babies might be switched.

One very good example is the use of RFID wristbands to track surgical patients in Birmingham Heart Hospital in UK. They are treating a very large number of patients every year and to prevent error in treatment, they are using this system for their thoracic and EENT wards. The wristbands issued to patients have an inlay of data with their hospital records and history.

Some of the NYSE RFID

A new technology for better business has risen. That is the Radio-frequency identification or RFID. Even the world's largest stock exchange, the New York Stock Exchange or NYSE has deal with using the RFID technology in some of its listed companies. That refers to the NYSE RFID or the use of the RFID technology of the companies involved in the NYSE.

There is no wonder why companies listed in the NYSE have decided to enjoy the benefits of using the RFID technology, most especially using the RFID tags. Through the RFID, different business organizations can ensure proper and accurate inventory of all their products. This means, they are assured that all their products are not stolen by any employee and that all of these will provide them better profit. Also, having RFID in any NYSE listed corporations will enable businesses to save more on expenses and have more productive employees.

Included to the NYSE RFID companies are the Clear-Count Medical Solutions, Intermec, Inc., and Texas Instrument Incorporated. The ClearCount is a medical device company which focuses on the patient's safety through various solutions. Meanwhile, the Intermec is involved in designing, developing, selling, and manufacturing different IT-related products. The Texas Instruments, on the other hand, is in for production of different semiconductors. All these companies are only few of the many business establishments which decided to integrate RFID technology in their businesses. There are still more NYSE RFID companies there. However, the stock exchange has no detailed list of all their listed companies which use the RFID.

Omega Engineering in the RFID Field

In the field of radio frequency identification, the company Omega Engineering is able to play a highly important part. The company has grown from simply a manufacturing one line of thermocouples in order to set the standard of becoming a global leader in this technical marketplace. It also offers at least one hundred thousand top of the line products for measuring and controlling temperature, pressure, humidity, strain, flow, force, level, pH level and conductivity.

Omega has been able to provide its customers with a wide range of electric heating, data acquisition and custom-engineering products. In the field of radio frequency identification, there are a lot of low cost features of Omega engineering that are really advanced compared to the more competitive brands. For example, Omega has a wide range of wind tunnels that can serve as a vane or a hot-wire anemometer. This is a lot more than what other companies can put out, to tell the truth.

Furthermore, their products are national and industry certified so one can be sure that it was able to get the seal of approval and rest easy in the fact that such products churned out by Omega engineering are of top notch quality. With the introduction of Omega engineering into the field of radio frequency identification, we can only be sure that only good things, better products and faster services are going to be laid out in the future. Clearly, this is good news for all and one can predict that more industries and companies will employ the use of radio frequency identification thanks to omega engineering.

The Birth of Patent RFID

The usefulness of RFID cannot be denied. The automatic identification system being used by the patent RFID has proven a vital source of data — both in terms of storing and retrieving it from a said source. By using various tags as well as transponders, the whole program is able to manipulate and transport data according to the data bank where it rightfully belongs. Moreover, the whole system is also not limited to one circumstance alone. It can be embedded safely onto a person as well as other living creatures such as animals. By utilizing radio waves, this patent RFID can finally gather the necessary information.

The whole tag is composed of a complex network of lines. Each tag is composed of two different integral parts. The other one is responsible for processing the information which gets retrieves and passes through the tag. This is made up of complex circuits inherent in its system. On the other hand, the other part is made up of an antenna which effectively transmits all necessary signals to the data bank. Chipless RFID is the process responsible for identifying the tags without the need the need for thorough identification made on each tag. In this way, they can be effectively printed out as assets at far lower costs than traditional means.

Because of the complex yet easy way it operates the patent RFID has been used by many businesses nowadays. Even government agencies worldwide have realized the usability of the entire process and has been using it for several modes.

A Safety Measure for Paediatrics

The growth of RFID technology has expanded and reached the level that it also provides control to hospitals as well. Controls to hospitals would include inventory handling, patients handling as well as security safety of patients.

Newborn babies and children confined in hospitals are the most helpless patients compared to elder patients. They cannot speak their minds yet because of their early age. A safety measure in most hospitals is installed to protect the safety of paediatrics. Hospitals have RFID patient locator applications. It becomes difficult for hospitals to track their patients. The use of RFID application is found to be a solution of keeping its patients on track and at the same time address its operational efficiencies. The RFID tag attached to a paediatric enables the system to scan the entry and exit points of patients and as to the location where the paediatric has to be at. It also provides information about the child and includes its health history as well. The use of RFID technology enables the monitoring of paediatrics without causing alarm to the rest of the hospital patients. The Paediatric RFID allows hospital personnel to avoid the mistake of interchanging babies and parents. Embedded RFID chips are attached to the paediatric wristband as soon as a newborn baby is born or a child is confined at the hospital. Children are a joy to everyone's family. As a provider of health service, don't you think that children safety is your priority as well.

The Advent of Radio Frequency RFID

The thing that makes the whole radio frequency RFID a phenomenon is the fact that it does not rely on cables of antenna to transmit data. It is a device which can easily treat data by simply emitting and receiving radio wav frequencies from other devices. Actually, most experts will say that the radio frequency RFID has already existed way back. However it was only limited to secret services back then and was classified as some form of confidential device.

But times do change and so technology brought with it the innovation that is known now as radio frequency RFID. It makes use of really strong radio waves to be able to recognize and identify personal information. It can also convert various signals and serial codes to a more recognizable platform for the end user. Because of the many different ways in which this technology can be used, both the private and government sectors have recognized the innovativeness of the said tool. Businesses are currently using the platform in controlling as well as protecting their confidential data. It also enables their employees to effectively retrieve data in a fast and timely manner whenever they need to.

Then of course, the entire platform has also become quite famous especially in the media. The entertainment and media communications industry have seen its potency in terms of utilizing potent data. Media agencies also ensure their quality control via RFID devices. The electromagnet waves which are key players for the said device makes it a worthwhile tool to use.

Retail of Radio RFID Tags

A lot of people who worry about the conspiracy theories behind the RFID tags technology do not in any way consider the more valuable aspects of implementing RFID technology. In fact, with the RFID tags, shipments of items have since been safer and more secure thanks to the RFID tags applied to them. Cargo containers these days use RFID tags to better monitor the cargo on board. With the high safety guaranteed for to be shipped items, shipping costs for everything will definitely be lowered. Efficiency of delivering products will also improve as chances of mishaps will decrease drastically.

RFID technology has proved itself so many times with big companies that even the US army uses these technologically advanced tags to monitor their equipment. The safety level that RFID brings for items tagged with are always highly considered by companies. And with today's technological innovations, greater enhancements to the way RFID trackers work are introduced which means better service.

RFID even considers using organic materials for its tags, as well as tags which no longer require batteries. The cost for RFID tags have definitely been going down. In fact, the RFID passive tags on a computer can now be simply printed out and it will still work. This sort of technology will definitely replace the old bar code system. If you really look at it, the way RFID technology is devised is not really for watching people and every move they make in a certain area, but rather it is developed for efficiency and safety measures which are so unrivalled.

A System Compatible for Device Net User

Your company's departments interconnect in order to come up with an output. It is in this process that communication among each department should take place. The exchange of data is necessary in order to relate actions required for a particular concern.

A communication protocol that automates industry to interconnect control devices for data exchange is called Device Net. Device Net uses Controller Area Network, the backbone technology. It defines application layer that covers a range of device profiles. Common applications include information exchange, safety devices, and large I/O control networks.

With the increasing demand for RFID in the industry, users of Device Net are able to apply their existing communication protocol with the RFID system. Communicator providers have come up with products that will allow user of Device Net to make their system work. This product is designed to conform with Device Net specifications and compatibility. The product acts as a slave device on the network. It presents serial data in the master controller, which in return processed easily the input/output data.

An RFID device compatible with Device Net provides operational flexibility. It supports IO sizes form 8 bytes up to 248 bytes. RFID system designed for Device Net has multiplex option that eliminates mutual interference challenges without the need for changing program designs. The RFID system for Device Net allows businesses to continue its existing protocol communication and same time provide quick access to information. RFID is ideal for inventory, part identification, routing and error tracking applications. It has features of graphical display and configuration keys that allow users for easy setup and operation.

The RFID Mouse is Wireless

Optical Mice is often powered through USB or PS/2 connectors or the use of AA batteries. The maximum life span of batteries for any wireless optical mouse is six months and shortest is three months depending on usage. Recharging batteries can be quite annoying as well as working on a corded mice.

An RFID Mouse is a device that gets its power through the RFID. It is a miniature sized tabletop RFID interrogator integrated into a computer mouse. It is designed to be handy to transport and use where RFID reading and encoding to stations be done with a laptop computer. The RFID Mouse has a mouse pad that has a cord and is connected to the computer via USB, which powers the device. The mouse pad is not cordless but the mouse is cordless. The mouse does not necessary require direct contact with the mouse pad in order to receive power. The mouse can be raised two inches off the pad and still get a signal. It can be used within a wide range of RFID applications in logistics, retail, supply-chain and other industrial uses. The RFID Mouse is commonly used in the programming and testing of transponders and labels, POS-applications, portable inventory management, and the control and identification of authorized users of easily accessible computers, terminals or other similar data entry devices. The Mouse is in UHF and HF versions.

The RFID Mouse Pad is a battery free wireless optical mouse. It is ultra light in weight and is portable. It creates huge cost savings, as you do not need to recharge batteries nor buy batteries, as it is battery free. Enjoy the use of your PC without having to be bugged with the cords being tangled up. Go wireless, go RFID Mouse.

What is RFID Supply Chain Management

It is every businesses primary objective to earn profit. In doing so, you need to be aware of how things are doing. If you are on the supply chain management, you need to keep track of the goods from the factory to store shelf. It is not only the loss of revenue that you should be concerned of but on finding ways to improve the productivity of transporting goods and securing the source of goods. The RFID technology is found to be the solution for supply chain management.

With the RFID technology, you can access on the 100% visibility of your warehouse products through the Supply Chain with RFID tags. The RFID is believed to be the ultimate Supply Chain solution that will bring billions of savings all over the supply chain and other areas. The RFID for Supply Chain Management (SCM) provides transparency in supply chain by cut out-of-stocks, counterfeit and shrinkage. One of the most important features of the RFID supply chain management system is the work force saving. The system is flexible to all kinds of applications.

Companies can do production planning, flow and process management, inventory management, customer delivery, after-sales support and service and trucks and ships tracking systems with the RFID SCM technology.

It may be costly to start on RFID, but the benefits of having it would yield a rate of good return of your revenue. It pays to put a price on a good investment. Keep track of your goods, and take charge.

RFID Security in Textile Supply Chain

Textile supply chains are a type of customers which RFID would cater to perfectly. RFID has been able to meet the expectations of such companies' needs by implementing further developments in how their software works. With these supply chains, computing expenditures, watching over inventory, as well as improving customer satisfaction rates is high priority. That is why they opt to choose RFID technology because with RFID software, the retailers can have complete knowledge of store operations such as stock information and the location of inventory in facilities. This system not only improves the way customers are served better, but also adds extra security capacity for the retailers.

All commodities within the facilities have been equipped with RFID intelligent tags which can be used to confirm inventory data whenever needed. Confirmation also only takes a few moments, without even requiring any sort of human requirement. With the use of RFID technology, retailers are able to save a lot of money with these. RFID also offers customers with added security for their products. Improved visibility can be initiated by the integrated RFID tags on goods. This technology decreased the amount of labour force needed to perform tasks, but an increase in sales is in line as well.

With the RFID tags, items are tagged uniquely. Tracking for these items has never been easier with the use of RFID technology, as well as the heightened security measures. Pilfering of items will be reduced with this technology by recognizing changes and fabrication of items. This is truly essential software for any supply chain company if they are concerned about their goods wellbeing.

What is a RFID Tag

The Radio Frequency Identification (RFID) is the best source of identification and tracking in today's world. It provides an automatic identification that makes use of RFID tags and readers.

An RFID tag is an object applied to or embedded into a product, animal, or person for identification purposes through the use of radio waves. RFID tags can be read even without direct contact and beyond the line of sight of the reader. It is resistant to paint and dirt. Depending on the application, a RFID tag carries not more than 2kb of data, a size enough to store information of an item. The RFID tags have at least two parts. The integrated circuit that stores and process information, modulates and demodulates a radio frequency and the antenna that receives and transmits the signal. The tags have tiny antennae that allows an RFID reader or receiver to receive the signal.

The RFID tags are of two types, the Active Tag and Passive Tag. The Active tag has a power supply that can receive signal from greater distance. It uses batteries that can last for a maximum of 10 years. The Passive tag does not have a power supply. It receives power supply from an antenna that produces incoming radio frequency scan. The passive tag can send and receive information and data over a short distance.

The RFID tag is used as an anti theft device. It is installed in most bookstores and CD shops. RFID tags will replace the UPC or EAN barcodes. The RFID tag is seen as an answer to keep track of things faster and same time increased productivity.

Understanding RFID Tags and Their Use

In order to make a Radio Frequency Identification (RFID) system work is to have an RFID tag. RFID provides automatic identification based on the stored data as embedded on the RFID tag. With the RFID tag, you are able to retrieve data on things bearing a RFID tag.

An RFID tag is an object, which can be placed to or included into a product, animal, or person for identification purposes using radio waves. These tags can be read several meters away and does not require contact. It can also be read beyond the line of sight of the reader. With this system data can be read through the human body, clothing and non-metallic materials.

RFID tags consists of at least two parts. First part is the integrated circuit that stores and process information, modulates and demodulates a radio frequency signal and specialized functions. The second part of an RFID tag is the antenna that receives and transmits the signal.

These RFID tags also come in general varieties. These are the following: 1. Passive Tag – tag that does not require internal power source. This operating power is generated from the reader. These tags weigh lighter than active tags, less expensive, and offer unlimited operational lifetime. They have shorter read ranges and require a higher-powered reader. 2. Active Tag- is a tag that requires power source. It makes use of an internal battery and is usually read/write. Example is when someone wishes to rewrite or modify an already tagged data. An active tag memory size varies depending on the application system requirements. Some applications operate with up to 1MB of memory. The active tag being battery-supplied power would provide a longer read range. 3. Semi-passive tag – tag similar to active tag. It has its own power source, however the battery only powers the microchip and no signal is broadcasted. Semi- passive tags have greater sensitivity than passive tags, it has better battery life span than active tags and can function under its own power, without reader being present.

What Is RFID

A new technology that is use in most businesses and organizations is the Radio Frequency Identification (RFID). RFID is a technology that provides automatic identification of an object, animal or person with an RFID tag. With the RFID tag, users are able to store and remotely retrieve data easily and quickly. The device is seen by enterprises involved in supply chain management as an aid in improving their efficiency in terms of inventory tracking and management.

RFID (radio frequency identification) is a technology that includes the use of electromagnetic or electrostatic coupling in the radio frequency (RF) portion of the electromagnetic spectrum that allows identification of things. The demand for RFID is increasing as it is seen to be an alternative to the use of the bar code. RFID is also called as a Dedicated Short Range Communication (DSRC) as it does not need direct contact or line of sight scanning unlike the bar code. Data is read through human body, clothing and non-metallic materials.

The RFID system consists of three components namely the antenna, transceiver and transponder. The antenna uses a radio frequency waves that provides transmittal of signal to activate the transponder. Upon activation, the tag transmits data back to the antenna. The data is used to signal a programmable logic controller of an action that has occurred.

RFID is around us. EZPass uses RFID as you go through a toll booth or on SpeedPass as you pay for your gas. You could just imagine if everything passes through an RFID system, things would be quicker and easier.

RFID Active Temperature Sensor: Intelligent Solution for Product Monitoring and Preservation

An RFID active temperature sensor is probably one of the most sophisticated and technology intensive varieties of RFID tags. The miniature tags serve as intelligent micro processors that can detect temperature changes in the environment and transmit this to a central database and monitoring system capable of triggering alerts. The technology uses active RFID gadgetry which means it has its own power supply making it possible for the system to work continuously. It has a sensor specifically designed to monitor temperatures of its immediate environment. It can be programmed to continuously monitor certain temperature levels and when a deviation occurs, it will automatically transmit an alarm for the said deviation.

Active RFID temperature sensors are popularly used in transportation, animal conservation, warehousing, and other activities needing constant and accurate temperature monitoring. The technology is very useful for transporting sensitive products like pharmaceuticals and biological products. The closed container vans of these products will be monitored 24/7 by the RFID tag while in transit. Data is constantly transmitted to a central monitoring system. Whenever a drop in temperature occurs, certain warning mechanisms will be triggered allowing authorities or handlers of the products to correct the anomaly. This technology is critical to preserve the integrity of products needing a controlled environment condition.

RFID temperature sensors are also being used widely in animal conservation and monitoring. This allows conservationist to practically determine the health conditions of animals at any given time. If a problem occurs, the RFID sensor will transmit data allowing authorities to act immediately.

Getting to the Basics of RFID

RFID or radio frequency identification is a method of using radio waves in order to identify certain objects. If you want to get down to its basics, this has been around for quite some time but is only now getting to be a lot more appreciated than ordinary bar code scanning devices. A radio frequency identification is a lot better because it can even track down and monitor an item without even needing to be in contact with it. To get down to the basics of the matter, the RFID can identify the stored serial number in a microchip that is attached to its antenna. Together, you call this a tag or a transponder. When combined, the chip can then transmit all identifying information going to the receiver. Thus, the reader will be able to convert any pertinent information into a digital format which will then be read by particular computers.

As previously mentioned, this technology has been around for some time, specifically in the eighties. It has been used in many toll booths, security access badges and automotive ignitions. Many businesses also use such a system in order to track down some products all throughout the entire manufacturing process, starting from the conveyor belt all the way to the packaging section. Such items may also be tracked and monitored while they are being shipped and received and even when simply sitting on store shelves. Clearly, it is very useful to have such technology around as it is able to make operations very efficient and successful.

These items can also be tracked while being shipped, received, and even sitting on the store shelves.

Basics of the RFID technology

RFID or radio frequency identification is an automatic identification technique to capture the data stored in an RFID tag (also known as transponder). Data from the tag are normally retrieved by a radio frequency (RF) reader through radio waves. The method is almost parallel with the bar code device however, instead of optically scanning the bar codes on a label; radio waves are rather used to retrieve data from a tag.

An RFID tag is any object that can be recognized in an RFID system. The tag is usually put on or integrated into a product, an animal or individual to function as identification through radio waves. A tag can either be read several distances away from the reader or even beyond the reader's line of vision. Moreover, a tag comes in various sizes and with different features.

On the other hand, an RF reader is both a radio frequency transmitter and receiver. It is now currently referred to as an RFID interrogator. The RF reader has an attached antenna that captures the data from the tag and then passes the data to a machine for processing. A reader can be used in a fixed position (e.g., installed on dock doors in warehouses), portable (e.g., incorporated into a mobile machine that is also used for bar code scanning), or embedded in an electronic equipment (e.g., labels of print-on-demand printers).

The RFID technology solution is commonly used among enterprise supply chain management (SCM). Basically, RFID technology can help in reducing warehouse and distribution labour costs; reduce the need for an inventory; reduce costs for point-of-sale labour; reduce, if not avoid occurrence of theft; and reduce some out-of stock conditions. Moreover, RFID adoption can also be useful in improving the supply chain's forecasting and planning and customers' shopping experience.

The Convenience of RFID Cards

One very popular thing nowadays is credit card that is powered by radio frequency identification. These cards are so in demand because they will allow you to purchase things using your credit card in a unique way – without having to enter even a pin number in order to do so. You do not even have to swipe your card on a reader or write your signature on the sale slip. What you do is wait for the coil transmitter found inside your card to send out a tagged radio signature. This tagged radio signature will get transformed by the motions made by your fingers over your actual card. Putting it simply, you may do the away with the card-swiping motion by using your fingers which is a whole lot cooler and not to mention a lot convenient for both purchaser and cashier. Imagine that it is also a lot harder for people to swipe your card and use it to make illegal purchases, because you and only you can use this card thanks to your electronic fingerprint.

There are many credit card companies looking into this integration of radio frequency identification. It is so efficient because you can use this at other areas such as gas stations, vending machines, convenience stores without worrying about bringing any solid cash or someone snatching your purse. Clearly, it is a good way to make shopping for your daily essentials and indulgence a whole lot easier and worry-free -which is all thanks to the convenience brought about by radio frequency identification.

The Silent Way of Identification: RFID Chips

Barcodes are very much a part of our daily lives that we don't really pay much attention to them. Ever cared to look at the barcode in one of the products you use at home? Let's take a look at example at the shampoo that you use. At the back of the bottle you can usually see a barcode with some serial numbers on it. When you purchased that, the POS system of the grocery just uses a scanner that is attached to the POS and voila! The name of the item and price comes out. Interesting isn't it?

So it was natural that a higher technology will eventually emerge to slowly replace the use of barcodes and that is the use of RFID chips. The use of RFID was first introduced way back 1969. Five years later, the patent was issued for it. RFID chips are very small, but can do a lot! For example, it listens to radio signals being sent by "RFID readers". The radio signal serves as wake up signal that request an answer. So we can see that RFID do not really need batteries to operate.

Most of the chips have radio signals that can be read from an inch or a greater distance, depending of course on the driving power of the RFID reader and the size of the antenna. The RFID chip costs about 50 cents, with the prices slowly dropping as more companies are manufacturing it. Since it is now cheaper than in the past, RFID chips are widely used in almost every business that you can think of.

For example, some governments are embedding RFID chips in their money to prevent the counterfeit parts and for identification purposes. Some pets are also embedded with RFID chips to help the pet owners identify them.

RFID: The Wonder Made by Chip Manufacturers

Have you ever heard of RFID or radio frequency identification? This is widely used by many industries all over the world and it traces its roots to the same tools that are used and handled by many a chip manufacturer. It was invented in the year 1969 but is now only gaining fast popularity in these troubled times. Chip manufacturers use microchips in order to make the radio frequency identification tags, and the smaller they are the better it is.

Some radio frequency identification chips are only one third of a millimetre in width. These objects made by chip manufacturers act as transmitters or responders that re always listening for important radio signals that are sent by transceivers or otherwise known as radio frequency identification readers. When a certain transponder gets to receive a particular radio query, it will be able to respond back by transmitting out its own unique identification code (usually one that is 128-bit) and goes back to the transceiver. A lot of these RFID tags that are manufactured by the chip manufacturers do not even have batteries but are instead powered by radio signals themselves, which signals them to wake up and request for an answer.

Microchips Used by RFID

RFID has already taken another step forward in aiding services such as supply chain providers by introducing RFID transponder microchips. These RFID chips are intended to be applied onto items used in a supply chain. These transponder chips are like the intelligent RFID tags which are used for locating certain items in a facility, which is in this case, a supply chain facility. These tags contain data about the item it is attached onto, depending on the item the data capacity is increased, which are read by the RFID system locator, used by the supply chain.

However, more and more RFID tags are being made such that they no longer contain any microchips. One reason is because of the demand for more natural organic alternatives when it comes to the materials used for the chips themselves. These are less expensive compared to the old chips which still used copper and silicon materials for production. Further development into these tags such as the anti-collision feature software, have also been integrated. With this anti-collision system, simultaneous readings of multiple items are made possible, and are read with ease.

These tags are either applied to the package of a product, or onto the product case itself. They are not noticeable as well, proving to be easy addition with a product package. The same tags have also been implemented in key chains for different uses, as well as for keyless entry methods with automobiles. All this software basically needs is to transmit the data the other point of the RFID system. Definitely the future software of utilities in the industry is the RFID technology.

RFID Companies: Providing Cutting Edge RFID Solutions for Enterprises

There are many RFID companies offering different radio frequency identifications solutions. The RFID technology is designed to effectively and accurately manage the database systems of companies in real time. Specifically, the technology provides companies with an effective control, monitoring and storage of their data needs for optimum corporate operations. However, not all RFID companies offer similar products. Each RFID company has a core competency and offers specific technology solutions designed for a very specific purpose. So it would be wise for companies to determine their specific needs and apply the appropriate RFID solution for their needs.

There are RFID companies that offer solutions designed for supply chain management. Products of these RFID providers could help companies in optimizing their supply management processes and supply chain tracking, monitoring and inventories. These could build up the capabilities of companies in efficiently managing their sources of important raw materials and distribution centres that are critical for the entire corporate strategy forecasting.

Other RFID companies on the other hand are specializing on the other end of supply chain management. Specifically, this refers to efficient warehousing, stock inventory control, monitoring of distribution processes, and retail management. The RFID technology can unify these complex operations and simplify the data needs for these functions. Through RFID technology, companies can be assured that their inventories and distribution services are in optimum shape and performance.

There are still other RFID solutions offered by numerous providers of radio frequency identification products. The most important thing that companies must do is to accurately determine what areas of their operations need the RFID technology.

Knowing Common RFID Company Failure

Choosing the best RFID Company could be a big headache for most CEO's or COO's. With so many RFID companies offering different solutions and services, it would take a great deal of time to find the right solution provider. Companies needing an RFID solution may look into some of the woes that could result if a wrong RFID solutions provider will be chosen.

First, some RFID products like tags, wristbands, security patches and monitoring labels do not conform to a standardized quality. For example, companies may discover that some tags are not transmitting data while others read better than other tag. There are quality issues that companies should look into before purchasing an RFID solution. The best way to deal with this problem is to require suppliers to conduct random product testing and to stipulate guarantees that have a concomitant financial penalty. The restrictions could ensure that RFID companies will only deliver quality standardized products.

Second, some RFID companies could fail to deliver the products on the agreed date. This is a common woe being voiced out by some companies needing RFID solutions. Delivery failures may disrupt the entire productions processes of companies resulting to significant losses of opportunities. To avoid this, companies must include in the purchase contract guaranteed delivery of RFID products and solutions. Failure to do so could result to contract termination at no cost to the recipient company. There should also have a financial penalty stipulation for failing to deliver the RFID products on the agreed time.

Advanced Computer Hardware RFID Technology

Most market forces these days underestimate the capabilities of the RFID technology. Most of them would only use the software for simple compliance tool for their supplies. They do not consider the other benefits of the RFID technology, mainly: an automated technology for assessing identities of individuals who use the system. The RFID technology in any RFID hardware is actually designed to increase a manufacturers production rates while lowering the chances of shrinkage in the processes. These are only additional benefits, but already they are very rewarding for any company to make use of, especially supply chains.

Only those with sufficient knowledge about using the RFID technology's fullest capacities and applications are able to turn around any company with the use of RFID software. That is why packaging suppliers with this amount of knowledge with evolving a company's RFID technology should be the ones sought after, especially if the company hiring the packaging supplier wants to make use of RFID tags.

Any RFID can be improved again and again provided that different combinations of software as well as hardware are used. The results are almost infinite, and the more efficient the results are, the better for any company. Dealing with supply and inventory management has never been much easier than with the use of the flexible RFID computer hardware. From intelligent tags which do not depend on batteries and are chip less to the Traceless® taggant by Creo, the latest in RFID tags technology, tracking items have been easier and further development and improvements are to be expected.

Steps to Ensure a Trouble-free RFID Data Management

There are several issues that may arise once companies deploy RFID systems to their operations. These issues need to be addressed immediately in order to avoid future problems that may result due to RFID implementations. The top issue hounding RFID implementation is how to handle the sheer volume of data that will be generated by tagging individual items. Enormous data that will be generated through RFID deployments could clog the databases of companies resulting to more problems for the enterprise.

Some experts believe that this could be solved by implementing a lean RFID data management solution. This means that companies need to take several necessary steps to ensure that their current data systems and databases will not be inundated by data coming from RFID devices.

One good measure that companies should undertake is to determine what data should be collected from the RFID systems and stored to its databases. Determining at first instance the necessary data that will be collected would solve half of the problem. Next, companies must install the necessary filter and buffering control mechanisms to sift the data transmitted by the RFID transponders. This will effectively gate keep information so that enterprise databases will not be flooded by peripheral or even unnecessary data collected by the RFID devices.

Finally, companies need to set the security parameters of their RFID data management systems to ensure a bullet proof data collection and gathering. This will also address problems related to data ownership. If these issues could be resolved, companies can be assured that their RFID deployments will certainly benefit the enterprise.

Data Mining: An Evolving Technology in the arena of RFID

Just like when any device or system that is geared towards the process of identifying the validity of any information, the need to employ the concept of data mining is inevitable and thus found to be very necessary. Data mining is the method of digging or a lot safer to say, extracting from hundreds of thousands of combined doubtful and legal information the information that is relevant and necessary. This whole idea is what is needed by radio frequency identification solely because the purpose of both technologies is all about securely identifying the legitimate information.

Data mining is very much reliant on the utility of the data industry. Similarly, because the origins of data may come from a very inconsistent mechanism and unknown and irrelevant relationships, the data that is used is highly vulnerable to potential attacks from malicious users. This is the reason why the concept of data mining in the RFID has been bombarded with so many issues on security and privacy. The privacy factor is a widely debatable topic among various sectors of the society. This is so because any person would be so much concerned about their own information privacy and that any leak on their information would mean an invasion of their own secure and private life. More so when you know that these pieces of information can actually jeopardize the life of a person.

Data mining employed in the RFID technology is not a new concept at all. In fact, it is not even a newer technology. However, the issue that revolves around it has been circulating so long.

Knowing the RFID Equipment

The RFID equipment is actually further divided into three different types. Each of these types or varieties has its own features and can also cater to different needs by various industries. In an effort to cater to a vast market, RFID developers have tweaked the tags to make them more flexible. The first variety is known as the passive tag. This particular RFID equipment is not dependent on any power supply embedded within it. It only relies on the electrical current being passed on to it via the antenna attached to the tag. This antenna receives some radio frequency and triggers the response which is needed by the system.

The opposite of this RFID equipment is the active one. Unlike the former, the active tag relies on its own power source and does not just allow its antenna to generate electric currents. In this way, this variety then becomes more reliable because it can process data in a timelier manner. It can also handle bigger sets of data because it has its own mini technological capabilities. The device can definitely thrive in various conditions like being submerged in water or even being included in ruthless situations such as animal herd conditions.

Then the last one is known as the semi-passive tag. This variety has a bit of both types previously mentioned. Although it has the same battery-operated life as an active tag, the power is only limited on the microchip and cannot be used in generating signals. In effect, the passive feature is thus displayed to the reader.

Sound Advice for RFID Implementation

It is very important to know how to implement radio frequency identification. The first thing you need to do is to establish your objectives clearly. What you ought to focus in is to start your radio frequency identification project with good and sound objectives. Whether you are just simply trying to comply with the demands of the retailer or trying to deploy radio frequency identification in order to meet the asses tracking needs of your clients, you will really have to make sure you get to draft out your real desired outcomes. Doing so will allow you to create a framework that will help guide your decision making.

Making trade off decisions will really come out during the entire process design of radio frequency identification and software selection, as well as radio frequency identification hardware and the scope of solutions. Having a clear set of objectives will really help to keep your entire team in tune to what is truly important and focus on the elements that you need to concentrate on.

Your next step ought to be to tag the entire system properly, or focus on tag and reader communication. If your reader will not be able to communicate with your tag, then your radio frequency identification system will not be able to work properly even if you were able to get really good software and processes. Your tags must be able to work by successfully transmitting data onto the readers in order for the radio frequency identification solution to operate properly. Then again, you also have to know that the effectiveness of both tag and reader communication is also governed by the physics of radio frequency. While a lot of implementers will employ several trial and error sessions by waving around a tag and asking if you can see it now, this will usually lead to an assumption of having poorly performing system.

RFID: Protection Against Counterfeit Parts In Car Insurance

Car owners usually buy their cars with an insurance attached to it in cases of some accidents or things that cannot be avoided. This is for their protection and supposed to be for their best interest. Supposing that a driver met an accident on the road and his car had damages, the natural thing to do is try to claim insurance for it. After all, he paid additional money so that in cases like this he can file a claim and the insurance company can take care of paying for the repairs and for the new parts needed.

For a new car, the parts would all be brand new. In cases where in accidents make it inevitable but to replace, there is a danger for both the car owner and insurance company that the replacement parts being used for the damaged car may not be original. Thanks to RFID, the automotive and insurance industry are now following suit in the RFID bandwagon.

The motor parts manufacturers have started using an RFID system to help identify the original from the counterfeit parts in cases of accident-repair car parts. Prior to this, insurance companies had little control over the parts used by auto bodies in replacing parts for cars claiming for insurance. In many cases, they get the counterfeit but they paid the price of the original parts. This is a lose-lose situation for both the car owner and insurance company.

Because of RFID, the insurance and the car owners can now be sure that they are getting the original parts untended for the car part replacement.

Publishing the RFID Updates with the RFID Journal

The use of RFID or radio-frequency identification has now reached almost all over the world. That is why, publishing the important developments and the basics of this new technology are now very important. Thanks to RFID Journal. They have satisfied that information hunger of every person interested with the RFID and how it works.

The RFID Journal is the first independent media company in the world which focuses on the radio frequency and how it is applied in businesses. The publication is actually now on its sixth year of successful information dissemination of the different events involving the RFID technology. It started publishing in the year 2002 because the founders believe that the then new technology will evolve into a higher form of necessity for the humankind. Obviously, that belief lived up until these days when more and more businesses are using the RFID technology to do better business and make better profits.

The RFID Journal's mission is to help companies know how to utilize the different benefits offered by RFID. The publication has foreseen the need for information about RFID when every year, a different industry variety joins the team of those using the RFID technology. To name a few, there are the aircraft manufacturing, consumer packaged goods, consumer electronics, defence, retail, and homeland security. The wide-spread use has not only for identifying products but also for tracking goods that will surely help in counterfeiting necessary products like medicines.

The RFID Journal has witnessed the evolution of this technology in its six years of existence. People interested with the RFID and the publication can choose to have the RFID Journal in print or reach them online.

The Benefits of RFID Labels at the Warehouse

RFID stands for radio frequency identification. This is a method of identifying and even locating persons and objects through the radio wave technology. Many companies, enterprises and even government agencies have implemented the use of RFID labels so as to be more efficient in data and information collection.

What would happen is that with RFID labels on products, radio waves are transmitted to the antenna. And then the antenna would then transmit the information to the reader. Through the reader, data is sent to a remote or central computer for it to analyse and interpret the data into a usable form. For inventory purposes, the use of RFID technology makes work much easier and more efficient since there's no need for any human intervention. In effect there is easier and faster tracking process in the warehouse.

What RFID labels would really bring to the company is the ease of tracking inventory. Just when it has a difficulty tracking products in the warehouse or it has a problem with visibility, the RFID technology could assist the company by the implementation of auto visualization of products. As for savings, there are a lot that the company saves on. The first is labour. The company wouldn't need people scanning barcodes. Nor would they need people for added security and to supervise effective inventory maintenance. Administrative errors and internal theft would also be reduced.

These are just a few of the benefits that companies or any agency would get if they implement RFID labels. And the use for the RFID technology is not limited for the warehouse. Almost any enterprise and even government agencies can use the technology too.

A Short Talk on RFID Manufacturers

A lot of manufacturers of radio frequency identification also come from the consumer section as well as the governmental one. Many market forces of radio frequency identification mandates are from companies such as Target, FDA or the Food and Drug Administration, Wal-Mart and the United Stated Department of Defence, or DOD. They have many manufacturers as well as suppliers that focus on the radio frequency identification as simple compliance tool which then overlooks the significant and additional benefits of using automatic identification technology.

If you think about it, the design of radio frequency identification is able to make for a more superior chain efficiency as well as inventory management which will then result into an increase in the productivity and the accountability of matters, and will be able to decrease shrinkage and come up with a more robust system. Once there is a packaging supplier that is knowledgeable enough regarding matters related to radio frequency identification technology and applications, such will be a really valuable resource that will be able to help you maximize the return of investments thanks to the integration system of your radio frequency identification.

These radio frequency identification systems are also completely customizable with almost an infinite number of combinations of both hardware and software. Each of the system parts is composed of what one will call a transponder, a reader as well as software. Manufacturers made sure to insert a smart tag, which is a microchip radio frequency identification transponder that is less expensive and is a lot more organic in nature, thereby making it the better option when it comes to choosing what will work well with the system.

How to Find Trusted RFID Manufacturers

With so many RFID product manufacturers and RFID solutions suppliers, companies may find it a little difficult to choose which supplier offers the best products. Companies needing the technology can choose from a wide variety of products such as RFID wristbands, key chains, lock and security systems RFID, patches, RFID tags, readers, among others. Some have very trendy design which can be used to minimize customer wariness about remote monitoring while others are very small and discreet patches which could be attached to packages to monitor deliveries and work in progress.

RFID technology has grown tremendously over the years that manufacturers are trying to combine utility, cost effectiveness, and aesthetics. In this vibrant market, companies must be able to choose wisely by checking several critical areas that determine the trustworthiness and quality of RFID product suppliers.

First companies must check if the RFID solutions and product provider is a verified supplier. This can be done by going to the global RFID organization of trusted manufacturers and suppliers. Usually, manufacturers of RFID products are reviewed and evaluated by their peers. Product quality and secure deliveries are the top considerations for verification. A verified RFID manufacturer means that it has passed the rigorous scrutiny of its peers. Companies then could be assured that they can get quality and secure RFID products from these verified RFID manufacturers.

Next, companies can also look for big name manufacturers and suppliers of RFID products. Although these manufacturers produce specialized RFID solutions, companies can always find the right products offered by these manufacturers for their business needs.

A Discussion of RFID Middleware

Using radio frequency identification by Middleware will truly help you understand and make sense of the radio frequency identification tag reads. Though sometimes it may be a bit hard to make sense of what kind of reading a radio frequency identification of Middleware has. A lot of analysts and vendors also have different ideas on what radio frequency identification of Middleware really is and what it tries to accomplish. They usually ask if it is a software, a hardware, and if it executes applications or monitor and manage several devices, and where it will go in the future. Of course, it is important to answer such questions in order to appreciate the radio frequency identification of Middleware for the valuable resource that it is.

As a form of definition, the middleware of radio frequency identification can be applied to formatting, filtering and logic related to capturing tag data by a reader. This is done so that the data will be able to be processed using a software application. To make the definition a lot clearer, there are other third party software that can also perform the same functions as previously described and then loaded into the readers of radio frequency identification so that nothing stands in the way of readers and the application software. The use for this stems from the fact that a lot of manufacturers end up making dumb tags as well as dumb readers. So they have no other recourse but to resort to radio frequency identification middleware in order to turn the data into the much-needed information.

RFID News: More Companies Embracing RFID

The advantages of implementing RFID must have been felt by those who first adapted the technology. And this has become apparent to other companies that more and more have decided to try it. And because of the demand, more and more companies have also given these companies many solutions that not only serve their purpose but answer their concerns as well.

Metalcraft, for instance, came up with RFID Hard tag. What they developed is not just an ordinary RFID tag and it was not created just for ordinary tracking applications. This RFID tag was designed to survive submersion in water and even heavy impact. It can withstand exposure to certain chemicals. And it can also cope to an impact up to as hard or as heavy as 20 tons.

When it comes to security concerns, Atmel and Mexel have their own offering of a highly secure RFID kit. This kit allows RFID developers to employ the use of mutual authentication and encryption even if they don't have an in-depth knowledge of cryptography.

On the other hand, Heathrow airport is trying its own trial implementation. They are using RFID tags on passengers' baggage for easy tracking. And although more information can be stored in these tags, they hold passenger name and destination for now.

A residential home in South Vancouver is also trying RFID in tracking their vital equipment. Should their trial be successful they plan to implement it on tracking and keeping tabs on their dementia patients.

Again, the number of companies implementing RFID just seems to be growing in number. And because of some criticisms and concerns about the utilization of this technology many companies are also coming up with better solutions.

Various RFID Privacy Issues

The use of RFID or radio-frequency identification has helped a lot in terms of better business process and prevention of counterfeiting products. But some people are not convinced with the rising popularity of this new technology. Some are even against the use of RFID tags on products and other services. Worse, some consumers are even intensively campaigning against products that use the RFID technology. The primary concerns of these objections are the RFID privacy issues.

Some personalities who are actively campaigning against the RFID technology refer the device used for this as "spychips." Even the California State Senator Debra Bowen has commented that now it is possible that someone's underwear can report on his whereabouts.

There are two main concerns about the RFID privacy issues. One is that the buyer may not know that his or her bought item has a device capable of gathering information about his personal doings. Doing surveillance against him is highly possible even without his consent. The other concern on RFID privacy issue is that the tagged item that is paid with a person's credit card or loyalty card can be used to indirectly deduct items from that same person's account. Thus, buying tagged items promotes high-tech stealing without the person knowing it.

To this date, there are still active groups of individuals especially in Europe and United States of America that are advocating against the use of RFID technology. These group's protest to raise the RFID privacy issues include boycotting of products that use this technology and debating over the security and other downsides of the RFID technology.

The Variety of RFID Products

Lots of different industries—both government and private—have seen the reliability of using RFID products. Thus, they are being used in many different ways. The new age of gathering data has indeed been upped by this new technology. With the used of this chip, people are now transferring and receiving data in a timelier and more effective manner. Among the most important RFID products have been developed in aid of passports. Nowadays, going in and out of some countries entails more than just stamping on the person's passport much more the VISA. The tags make the entire immigrations transaction more valid and secure.

Another one of the most effective RFID products is using it as means of paying for transportation. All kinds of payment are made much faster and accurate through adapting the said system. The tag is being widely used in the European continent as well as in some Asian countries. Because of the said system, public transport operators no longer have to worry about freeloaders and they can securely focus on just transporting the passengers on their proper locations.

Lap scores in various racing contests also make use of tags. Because of high speed and other off-road challenges faced in the competition, only these tags can accurately record speed as well as detect any machine glitches during the race. Speaking of tough conditions, these RFID products are also being extended in the world of animal farming because herds are rather hard to follow individually by manual records.

Scrutinizing the RFID Reader

The use of the RFID or radio-frequency identification in the evolving technology that we have nowadays may not seem to be simple and in fact involves too many intricacies in terms of its design and ergonomics. The RFID is usually containing tags which is capable of being applied or to a more highly technical term "injected" onto a product or good and even to humans which is a prevalent identification system to most developed countries. This tag element is capable of being recognized thru the use of the equally intricate and highly sensitive RFID readers, usually extended beyond the reach of human sight.

The readers used by RFID are designed based on the function that it was made for. A smart card for example requires a different reader compared to a passport RFID. This is done to make sure that lesser complications are being felt with the use of this technology. A reader may scan an area in the tag in various modes. For instance, when the tag is in the mode of autonomy, the reader then will sporadically search and locate all the existing tags in the range of operations. In addition, in this mode, the reader will then safe keep a list indication of the presence of the tag along with the time of persistence it contains along with other control information. The reader then can identify the termination date of the tag and potentially remove these tags on its list.

All in all, the process of the reader may seem to look very complicated, however, once perfected, the whole mechanism is simply a work of a genius.

The RFID Reader: Learning and Understanding the Technology

Have you ever wondered how a mobile phone is actually "talking" with the SIM card that you insert on it? Or more practical to ponder on, have you ever asked yourself about how your ATM card is being read and re-transmitted with data back to the point where you inserted it? Wonder no more, because the answer to your question is very simple – it is with the use of the peripheral device known as the RFID reader.

The card or the chip that you use to store your bank or personal information is residing contains millions of small circuits on it that allow for it to be "asked a question" or talked to" by any compatible RFID reader. The reader has its own transmitting and receiving mechanism, usually in the form of the antenna that transmits radio form waves; the tag, a primary component of any RFID usually answers back by sending the data sent into the radio wave. Now, just like any principle of transmission and reception, there are several factors that greatly affect the quality of the data. The distance of the transmitted data from the receiving point plays one big role. The reader will find it a lot difficult to be transmitting and receiving the data when the distance that it travels is quite long. The heavy traffic during the transmission caused by interferences may also contribute in the poor reception and transmission of the data. However, all in all, these problems are rarely being experienced because of the quality of integrated circuits that are being used in developing these gadgets.

A wireless technology: RFID's unique usage of tags and reader

Radio frequency identification (RFID) is an automated identification system, wherein data integrated in an RFID tag or transponder is captured by a reader using radio waves.

Basically, an RFID system consists of RFID tags (or transponders) and reader. An RFID tag is any object that supposedly contains essential product information, specifically stored in microchips. Tags are usually attached or installed inside the inventory or equipment. In addition, RFID tags can be in various sizes and with diverse features.

The reader (also known as RFID interrogator), on the other hand, can be a standard Windows-based PC device. RFID reader would capture the data or information stored in these microchips by using radio waves, and transmit them into the computer. Moreover, the reader can be portable, implanted in an electronic device, or used in a still position.

Since RFID's initial development, RFID technologies have been progressively improved to cater to the increasing requirements among businesses and companies. Purposely, RFID systems were designed to use modern wireless technologies. RFID systems are ideal particularly for businesses to carry out business track inventory and equipment.

Compared to the traditional or other modern supply chain operations systems (barcode systems, for instance), RFID technology is rather commonly preferred by companies of various sizes and nature for affirmative reasons, which are: 1) no line of vision required; 2) long reading range; 3) tag used are guaranteed durable, thus can withstand severe environments; 4) use of portable database; 5) use of multiple tag read/write; and 6) capable of real-time tracking (e.g. items, people).

Generally, RFID is by itself a wireless technological approach that can help business enterprises to achieve an improved holistic supply network management.

RFID Research Brought out Many Great Things

Quite obviously, the development of RFID technology required so much research before it became fully functional. Although the beginnings of such development was traced back to be for other purposes, what is in use now has given so much benefits to companies implementing it. Among the first true RFID utilization were, for example, for toll systems, vehicle identification and electronic plate numbers. All of these are still in use today. The banking system was benefitted by this technology. RFID technology made the present electronic check book and electronic credit card system possible and manageable. Security systems also improved through the use of RFID in identifying personnel, automation of gates and surveillance. The medical sector, of course, was able to use it in tracking patients and storing patient information as well.

RFID research truly has brought about many uses for the technology. It helped companies to be cost effective. And as a result they were able to get better company performance. However, there have been many concerns and criticisms too. And these had to be answered if the technology was to evolve and to continue to be implemented. Again, through RFID research the technology was further improved. The first RFID tag, for instance, was a passive RFID tag. It was later on improved into an active RFID tag so stronger signals can be sent out when needed. RFID hard tags were also recently developed so as to answer the need for better RFID tags for problematic situations. These tags were now less susceptible to hard impacts and submersion into water.

There are still concerns left unanswered and there are still a lot of improvements to be done. But in time and through RFID research, things would even be better and more beneficial in the years to come.

The Business Behind the RFID Retail

In the business world like in a retail company, both customers and the management want to attain convenience in business. And as an entrepreneur, one may want to give the highest satisfaction possible to his customers, and the aftermath of the customer's fulfilment is the increase in business sales.

Radio-frequency identification (RFID) gives intimate pleasure to an entrepreneur. With RFID, business establishments and institutions will have a real time visibility of their product movement and inventory to have an assurance of store productivity and lesser cases of lost items. Inventory managers are always saving precious time in monitoring and controlling their supplies. Stores have cost-cut their expenses because they attain a bird's eye view of their stocks and their product levels, thus unnecessary orders will be eliminated. Store managers will be at ease in observing quick-selling items, and will have full control of the inventory supply.

Better business is a synonym for customer satisfaction. Through the wonder of RFID, the salesman or staff in a retail store has comfort in identifying the exact spots where the retail items are located. The requests of customers will be handled in no time because of a centralize database.

Due to the many benefits of RFID retail, more than 200,000 suppliers and manufacturers are now driving the global market for software and hardware to boost the use of this technology.

The future shoppers could also use the RFID technology. Through this, retailers will have the knowledge of the purchasing fashion of their customers. With the collected information, a retail business can know what their customer needs and they could devise strategies such as promos to boost their customer loyalty.

The Innovation of RFID Scanner

People say that the RFID scanner can actually replace the other laser technology out there. It brings the whole scanning thing into another level. This is because the RFID scanner is not as dependent on codes just as the usual laser barcode machines are. Although it is accepted that this new technology may not be able to totally replace the old barcodes due to higher costs, it is expected that they will bring the whole coding technology on a much higher level than it is currently in.

With the RFID scanner, data is being transported easily and recorded. However, the whole process takes a lot of method to undergo. It generally requires loads of terabytes in order for it to function properly. However, there is always a unique identity which plays a vital role in each RFID scanner. This can never be manipulated and multiplied. If it is not recognized then the entire data transfer will fail. In addition, the big data size capacity of RFID makes it more flexible in terms of motion. It also has an antenna inherent to its feature so it can be tracked easily and monitored without the need for the actual tag to be present. As long as the data of the tag has already been scanned or logged beforehand, tracking down the tag's presence no longer becomes a problem for the administrator.

In this manner, the RFID scanner will no longer need the typical laser machines which can be found associated with barcodes. Transfer of data is virtual and dependent on waves. Thus, interceptions are going to be lessened.

Questions about the RFID Security

There is no doubt that the RFID or radio-frequency identification technology has provided a lot of benefits for humans. Most of these are for businesses where they can have better productivity and better business. They can even have better inventory and better assessment of their goods. Not only that, RFID also helps in stopping counterfeiting of different products like watches, clothes, shoes, and medicines. However, some asks, "Is RFID really secure?" The remaining question that might put this new technology in the trash bin is all about the RFID Security.

The primary use of RFID technology is to provide unique identification for products. However, that same technology can also be the reason for anyone's unsecured living. One may buy a watch with the RFID tag or buy a pair of shoes with this same tag. What happens is that the tag will remain in that product even while the person is wearing them. Through that, some devious people can just locate anyone with the RFID tags. That makes the RFID security in question.

There are also concerns about identity theft due to the RFID technology. Some restaurants and bars use RFID tags for their VIP customers. Through that, they can easily get in the bar and pay their bills easily. However, that same RFID tag can be copied and be used to enjoy all the luxuries supposed to be for the original person. Privacy is also a big concern regarding the RFID technology. A certain US politician once commented that a person's underwear can just be the reason of tracking his location.

Because of these concerns, there are now individuals involving themselves in the fight against the use of RFID technology. These people believe that there is no RFID security when this technology still exists in many stores worldwide.

RFID Software: Getting More Lucrative as New RFID Markets Emerge

RFID software is the backend support for RFID hardware deployments. In other words, the RFID systems such as the chips, antenna, transponders, and the actual tags are the infrastructures while the software processes the data generated by RFID so that administrators and data or security handlers could read it. RFID is a standard device consisting of a data storage micro chip and miniature radio wave transmitters. It would be useless if its transmissions could not be read by end users. It will need software support so that its transmitted data and information would be useful.

Market analysts predict that the market for RFID software could increase by 1000% in two years. This is due to the fact that more companies are adapting RFID technologies for their enterprise data solution and management. RFID is also being used extensively to efficiently monitor the flow of goods and implementation of work in real time. That is why giant software developers and vendors are barging into the middleware market or acquiring start-up RFID software companies to get a chunk of the RFID market. The mergers and acquisitions could standardize software solutions for RFID technologies and bring about mass produced applications for the hardware. This will certainly create new markets thus driving up software sales in the retail sector.

As technology for RFID gets more sophisticated, software developers will also continue to roll out increasingly advanced software solutions. This will result to increasing efficiency of the whole RFID systems therefore more companies will certainly adapt the technology. This will boost sales of RFID software which could drive further growth of the RFID industry.

Ensuring Data Integrity through RFID Software Internet Security System

RFID hardware works like a mini processor. IT crunches data from the source and transmit it over radio signal to a terminal or RFID reader. The data then would be read by software capable of supporting this remote data transmission and filtering information based on the configuration of system administrators. Of course this technology, especially those using Internet protocols as data channels, could be subject to security hacks and data hijacking. That is why a parallel internet security system interfacing with the RFID software is necessary in order to secure transmissions and ownership of data would not be compromised.

There are many vendors of RFID software with Internet security features. This software can be used on standalone readers in order for the device to communicate with computers using the Internet. It ensures secure delivery of data increasing the confidence of companies that are using the RFID data management solution. It can also be utilized to enhance product tracking and the data transmitted can be directly channeled on Internet networks for ease of use and convenience.

One unique utility for RFID software with superior security systems is tracking live animals for conservation purposes. A conservation groups may tag individual animals and configure the system to transmit data over IP channels. In this way, tracking of animals can be done in real time with data being displayed on web interfaces. Securing this data transmission will be handled by the security software to ensure that accurate information will be delivered to the administrators.

RFID Software as the Leading Tracking Technology

These days, with the huge boom of much bigger projects in the world of industry, watching over the asset for the project, the important data and the people working with the project are essential. The need for better tracking systems have suddenly increased and RFID Software proves to be the most effective client. RFID software deals with remembering the valuable project information in place of humans. And no longer is it necessary to use the old lengthier data histories with any valuable data intended for any industry's project. It is because of this innovation that various similar tracking systems are available in the market, all competing against all the leading tracking technology with RFID in the lead.

The RFID innovation made a big impact in this field because of it is simpler yet more effective approach in dealing with tracking the data. Using radio signals, RFID no longer needs go through a whole history of the data but instead, only uses indicated tags for locating specific items. In a warehouse facility for example, locating packages have never been easier. With the use of the RFID intelligent tags, all one has to do to locate an item in the area is by scanning the RFID video tag.

The RFID software will soon hit the fields of supply chains, retailing companies, and manufacturers. This will surely be a big hit in the industry to watch out for. For now, RFID technology is the leading tracking software in the market and can be found at agilesense.com.

Good Solutions from RFID

In the realm of radio frequency identification, it is good news that the solution is able to add yet another chapter when it comes to asset tracking technology. These are times where a large volume of data is important and on can truly say that additional manpower and technology is needed to complete the corresponding big projects as well as good tracking systems. In this age, radio frequency identification solutions are necessary in order to make systems a lot more effective. The solution of providing asset tracking system will be able to help you when it comes to monitoring their important as well as valuable assets. This is the reason why such devices and such solutions are so in demand.

In radio frequency identification, the most ground breaking technology is upon your hands. Now, there is no need to store a big volume of history and information as well as the full description of them each and every time. Using radio frequency identification tags is a truly simple way to monitor them. Each tagged asset can easily be tracked by interpreting all the upcoming radio signals. These solutions are necessary to track down the particular servers that come with the entire packages that have to be located in the warehouse. There are also a lot of clients coming from the manufacturing industry that also make use of these radio frequency identification solutions such as supply chain management and the retail industry. A lot of vendors are excited about the arrival of the solution and are making the necessary moves to accommodate it in the systems.

Why Invest on RFID Stocks

Investing in RFID stocks is an excellent stock market investment move today. That's because the RFID technology sector and some individual RFID companies are posting bullish growth at the trading floors. The value of RFID stocks have been soaring considerably for the past months due to several favourable factors.

First, more and more companies now are embracing RFID technology for their data management solution and other business application needing an accurate identification processes. This event fuelled actual retail sales of RFID products as well as boosting the demand for RFID solutions. On a macroeconomic level, investments on the RFID sector have shown vigorous activities with expansions and breaching of new international markets. With the entry of giant corporations, specifically in printing, micro chip, and IT industries, to the RFID market, stocks begin to soar. That is why investors in the RFID sectors have experienced sudden windfall and more are coming as the technology gains more foothold in different industries.

Second, investors could predict long term RFID stocks growth and solid returns due to the increased government spending for RFID products and solutions. The state sector occupies a big chunk in the RFID consumer base because the government has been using the technology for its different departments and critical functions. Specifically, RFID solutions are increasingly being used in the defence, home land security, and trade departments. These sectors alone can significantly drive RFID sales which would definitely impact on the value of RFID stocks. So it would be wise for speculative and long term investors to funnel their attention to the RFID sector because it now holds the top position in terms of industry growth and expansion.

The Apparent for Need RFID Suppliers

Today the use of RFID technology has been gaining more and more popularity. More and more companies, enterprises and government agencies of various countries have started to implement the use of RFID whether it is for easy warehousing or tracking of people. As the name suggests, as long as there is a need for individual identification there is a use for RFID.

Most enterprises would use RFID to replace the use of barcodes. Many companies would even require their suppliers to implement RFID in their goods so as to have their own inventory have a uniform means of identification. And so these suppliers are then sometimes forced to implement such just so they can continue doing business.

Because of this the demand for RFID labelling and such has increased. With increase in demand, the number of RFID suppliers has increased as well. Most RFID suppliers would provide applications such as asset tracking. Companies would know when a certain expensive asset is being moved and where it is in an exact moment. Then of course, there's asset security. Important objects and assets of the companies cannot just be brought out by any personnel. Only authorized people can gain access to where the asset is and to the object itself. Yard management, too, can be done better with the help of RFID technology. With it in place, the coming in and out of products from certain premises is monitored better.

These are not the only applications which use RFID technology. And the RFID suppliers needed are not limited to those who can supply the applications and solutions. There's also a need for RFID suppliers for the equipment. But then it is not a problem because where there's a demand RFID suppliers would always be present.

The RFID System Goes More Steps Forward

The RFID technology system is taking its spectacular innovations several steps at a time. In the field of education, several schools in Osaka, Japan have been implementing RFID software for tracking the students. RFID chips are being placed in students' cloths, bags, and their identification cards. Schools in England have also implemented the same tracking devices onto student uniforms. This is another step in further improving the safety of a school facility's students.

Even in museums, the use of RFID ground breaking technology is visible. In museums, RFID works with education as cards integrated with the software are handed out to visitors, giving them information about the current exhibit which they are observing. The RFID integrated cards can even be used to take pictures which can be viewed at the museum's gift shop and bought.

Even in social retailing services, the use of RFID technology is not alien. In walk in store dressing rooms and with only the use of the RFID electronic tags on the clothing and an RFID antenna, viewing the apparel on the customer and on a celebrity at the same time is possible. A webcam inside the dressing also transfers images of customers wearing the pieces of clothing to a website which can be viewed by other customers.

RFID technology has certainly been making the best software for its customers and further improvements onto the system are sure to be expected. Right now it has hit even schools and retailers, what more in the near future?

The Usefulness of RFID Systems

One useful system nowadays is the radio frequency identification system. it is a highly advanced and automatic identification system that uses the Auto-ID system which is based on radio frequency identification technology. Why does this have a high value for many inventory systems? In particular, such a system will be able to provide a highly accurate knowledge of the current stocks or inventory. For example, in an academic study it showed that at Wal-mart, the radio frequency identification system was able to reduce the quantity of their out of stock products by over thirty percent for some products that usually sell between less than one percent units per shop day.

There are other benefits of using radio frequency identification systems as well, like a reduction of the cost of labour, being able to make business processes a lot simpler and reducing the number of inaccuracies in inventory.

Around four years ago, some companies decided to integrate the radio frequency identification system in order to help significantly reduce the costs of maintenance as well as inventory on aircraft industries. Because the costs of aircraft parts are quite high, the radio frequency identification technology allowed them to monitor each piece and thereby save on costs. No other system can help do this for them at least – or at the most make significant enough reductions and savings like radio frequency identification. It then comes as no surprise why it is very in demand nowadays and used in a wide variety of industries.

What are the Uses of RFID Tagging?

Radio frequency identification uses tags for the purpose of data storage, monitoring, and retrieval. RFID tags come in different sizes and shapes but the most commonly used are thin plasticized band containing a micro chip and transmitter. It is similar to tags being applied to baggage. However, there are RFID tags that are used on animals which are sometimes implanted superficially in the skin. RFID tagging is applied to accurately track the movement of entities usually for security and monitoring purposes or to determine actual work done specially on deliveries of goods.

RFID tagging is also widely in healthcare, hospital management, and patient monitoring. Patients will be tagged so that the medical records and information can be easily retrieved by health professionals. This will minimize medical malpractice and avoid incorrect administration of drugs and medications. Some pharmaceuticals companies are also RFID tagging their products to eliminate drug counterfeiting or theft.

In government services RFID tagging has been applied to tanks and jet fighters primarily to avoid friendly fire engagements and to monitor or track the positioning of stealth bombers and other aircrafts. Passports will also be tagged to enhance the security of airports and passenger airliners.

The retail sector also applies RFID tags to individual items to effectively monitor sales and determine market demands and consumer pattern. RFID tagging is currently being used by supermarket chains to efficiently monitor item supplies and inventories.

There are many uses of RFID tagging. It is a good technology to monitor the movement of products. It can also be used as a security measure to strengthen current identification system.

Better Management with RFID Tags and Readers

Is it possible to construct highly robotic or technological warehouses for industry giants such as Wal-Mart, Gillette or even our very own Department of Defence? Is it possible to make these systems run on a more efficient level in order to save energy and reduce costs? Can there be a system that will help such companies make automatic motions on all their things and equipment so that there will be no need for additional manual labour, such as turning off lights and electric systems in the event of natural disasters? Clearly, the need for a system is evident but we are lucky that such a technology called radio frequency identification is present in order to help us fix all these things and prevent future worries.

Many companies are now testing out how well RFID tags and readers work for these systems and so much more. If you combine these radio frequency identification tags and readers with conveyor belts and the robotic forklifts found in many a giant companies' warehouse, the electromagnetic induction technology will be enough to charge up a capacitor better than simply working the headlights of the power grid or even the battery of a back-up power system one used to use in an emergency. Using such robotic machines coupled with the technology of radio frequency identification will be able to let a lot of warehouse managers breathe easier. If the radio frequency identification-charged LED tag readers get a bit too dim, then they will simply revert back to back-up power.

RFID Technology Vs. Bar Coding

RFID technology has proven itself an important tool in the private and government sector. And the whole thing turning out to be this way is not really that surprising, considering the fact that it is a very potent tool which can function in more ways than one. It is proves to be a one of a kind system in terms of personal identification. People who use it for such purposes find a very safe and secure way of treating information. You cannot easily tweak its data unless duly authorized by the system administrator.

The whole RFID technology is reliant on radio waves and it can easily detect changes or interruptions. This RFID technology a whole lot different as compared to the usual bar code functionality. Although both of them employ the usual coding functions in order to transmit data, a bar code will simply utilize various optical signals in order to recognize codes and transmit the data. As for the RFID, it makes the whole process much more complex by using the radio frequency signals to transmit and identify necessary information. In essence, it becomes more secure because radio waves can be harder to change or intercept than bar codes.

In addition, the RFID technology is also more interactive. Bar codes are often used to transmit data to another host but it cannot transmit data between different car codes. RFID can do this, provided the recipient mechanism has an internal device which can actually read data based on the RF feeds.

RFID Test Equipment: Ensuring Quality and Performance of RFID Products

RFID test equipment is any device, electronic gadget, or software that determines if the RFID system or any component of it is working properly or not and can be used also for designing RFID devices. RFID test equipments are extensively used by manufacturers for quality control purposes and maintenance of their system.

Varieties of RFID test equipments range from software simulators, signal testers, electronic configuration analysers, among others. Integrated software systems and simulators are used for analysing circuit levels, electronic layout and design of RFID end products such as tags, plastic bands, and transceivers. These simulators can accurately measure the required circuitry and the optimum dimension of any RFID systems.

Aside from software design simulators, there are also RFID test equipments that analyse radio frequency strength and efficiency of signal transmissions. Most of these RFID analysers simulate planar electromagnetism of RFID circuits. It can measure the effects of electromagnetism on the performance of each RFID components. Analysers are very useful tool for determining if the electronic design and configuration of the RFID system is working fully. This ensures that RFID products rolling out from the production line will perform under any environments.

Another variety of RFID test equipments are signal generators and adjusters. As its name implies, it test-run the capability of the RFID transceivers to send and receive signals through calibrated radio frequency signal test. It is a powerful utility to ensure that the transmitting and receiving capabilities of RFID tags are working at maximum efficiency. It also test the ability of the system in handling signal interference which is a fairly common problem of any RFID product.

Some Issues Associated with RFID Tracking

RFID tracking has a chilling effect on some sectors especially consumer protection groups and civil libertarians. The main issue against RFID tracking is the decimation of individual privacy. Concerned groups assert that tagging individual items with micro chip is tantamount to peeking into the private homes of consumers. However this fear has been continually being addressed by RFID companies. Specifically, manufacturers of the technology assure consumers that the RFID tracking tags can be disabled by the consumers with a simple scissor. They could easily cut the tag therefore effectively short circuiting its data transmission.

Proponents of RFID tracking also point out the numerous benefits of electronically tagging an item. These include effective measure against counterfeiting of products because through RFID tracking, individual product package will have its own unique number code. If a counterfeit product does not possess this code, then the consumer and retailers can easily be alerted that the product is counterfeit. RFID tracking also has many security uses. As of now, the technology of RFID tracking will be implemented to effectively track movement of people in high risk areas such as airport terminals and airliners. RFID tracking will also be implemented to the passport system to enhance the security of the document. It will effectively cut fraud in this area making sure that important documents will not be used for unscrupulous and criminal activities.

RFID tracking, as with any other technology, has advantages and disadvantages. However, the enhanced business intelligence and security that will result due to RFID tracking will certainly benefit more people especially the consumers.

RFID Wristbands: Providing Convenience and Security for Enterprise Data Management

RFID wristbands are specially designed electronic tags used extensively to quickly and accurately store, track and retrieve data through the use radio frequency transmission. In fact RFID stands for radio frequency identification which implies the use of radio wave transmissions for managing data. RFID wristbands are actually part of an entire RFID solution that may be implemented in an enterprise, whether in manufacturing, distribution, or services. The RFID technology is a complete set of electronic data and retrieval management systems so that enterprises can simplify and effectively control their stock inventories, customer data, financials data solutions, and security. Aside from these, RFID wristbands can be applied to any data management needs of enterprises so that their database utilities could be systematized.

RFID wristbands are lightweight, razor thin arm bands equipped with micro chip for data storage and an antenna for remote transmission and frequency receiving. It is a cutting edge technology but it should be noted that RFID technologies have been in the market for quite a while now. It is just now that its importance and utility has been getting wide attention and companies are increasingly transferring their data management monitoring to RFID solutions.

Some of the best RFID wristband can be used over and over allowing for resetting and reprogramming of its data storage function. Some on the other hand are one time use RFID wristbands which are utilized for specialized purposes. Enough security has been installed on each RFID wristbands in order to avoid unauthorized data retrieval or hijacking. Most RFID wristbands are also tamper proof through the use of special circuitry.

The Security Implications of RFID

Just when we are thinking about evolution and globalization, the security issue becomes a real hot topic. The use of the RFID in all sectors of the society, from the world of medicine to the school sector down to navigation and commerce, the threat to security is becoming a compounded element of discussion.

It has been said by many experts that radio frequency identification is a highly secure technology. However, the issue that surrounds this technology in terms of illegal monitoring and tracking of the tags used in the RFID has not ceased and has increasingly become very alarming. Even the highly developed countries that use RFID as their major tracking and monitoring systems technology is fully cognizant of this threat in the RFID tags.

So, how does one prevent their RFIDs from being cracked and hacked?

One of the more common security mechanisms used by majority of the users is the use of encryption methodology. This method allows the tags to be placed on a higher level of being secured in areas such as being cloned, hacked, or cracked. Another more useful and evidently secure method is the use of the dynamic responses or potential pattern to crack the tags every time it is being scanned. This method allows for the tags to change every time that it is being scanned thereby, making it difficult for potential "cloners" to imitate and copy the tags.

The issue of security is a going to be a long running debate. However, the good news is that, the RFID to this date remains to be still secure and effective in all aspects.

RFID technology: some shipping benefits

By definition, RFID or radio frequency identification is an automatic identification system by which data in an RFID tag or transponder is captured using a radio frequency (RI) reader. An RFID tag, which usually comes in various size and features, is an object that holds data. An RI reader (or RFID interrogator), in contrast, is the device that uses radio waves to capture the data integrated in the object, and transmit the data into a specified machine. Since its development, RFID technology has been commonly adopted in various enterprise supply chain management (SCM) for its numerous benefits.

Specifically, RFID is remarked to help improve both visibility and security of shipments among the biggest importers and exporters. Basically, RFID gives these companies a certain level of security to their shipping containers. It provides shipping companies competence in securing their shipments since they are able to record and detect any occurrence of shipment interference. RFID thus offers essential improvements in the tracking and monitoring of shipment transactions. Also, RFID technology provides an occurrence of errors in shipping operations by guaranteeing that all fulfilled items are completely secured before they are dispatched. Moreover, RFID can also afford significant gains output among large shipping companies since it allows shipments to be classified as low risks, which results to much faster inspection thus, an instant profit is achieved by them. Lastly, RFID not only offers security in shipping operations, but can also provide considerable cost savings, especially when carried out with some tracking network.

In conclusion, the RFID technology continues to prove it potential as an advance solution to further improve various companies' supply chain management all together.

Revolutionizing the Globe with Smart Card RFID

In this time of global challenge, the need to become innovative, expeditious, secure, and high-tech in all aspects has become more of a pre-requisite to survive the battle in global economy than just being on-top of everyone else. Now, every corner you turn your head to, the use of the Integrated Circuits embedded on a card to either to transact business or merely just to safe keep your information, is very much evident.

The Smart Card RFID is one revolutionizing technology that a lot of businessmen and technology developers are geared towards into tapping in its utmost potential. The Smart Card RFID is a minute, small-sized card which carries on it millions of small circuits capable of storing, retrieving, and manipulating hundreds of thousands of information. This is fast becoming not only the trend but a necessity to all. The Smart Card RFID is usually characterized by a highly secure, hacker/tamper-resistant which contains hardly encrypted codes to maintain security and confidentiality of the information saved on it. These smart card RFIDs, just like any other smart cards make use of an equally secure reader that allows for the transmission and reception of information or similarly the exchange of information from one point to another.

Now, as we turn a great leap, the smart card RFID is the ultimate manifestation of us, being ready and adaptive to a more effective, more conducive, more flexible, and a lot secure business and day-to-day transactions that requires little amount of intervention from humans. This spells great challenge but with great comfort when achieved and realized.

RFID and the Software Requirement

The concept of RFID is a tremendous leap that human brains have come to perceive. This is one major evolution that human kind has learned and developed in full potential. However, the use of such technology is usually engulfed with issues of security, dependability, and reliability. This is where the secure layer is being tackled and discussed in a more detailed perspective. More often, the employment of encryption method to maintain a highly secure environment is made possible with integration of software engineering. Every device that works on a highly advanced technology is more often associated to run and execute with the help of some perfectly designed software. RFID is by no means an exception.

The RFID software is a specially designed software that particularly works with special types of RFID device. The sensor for a "bar code symbol" requires a specifically designed software that works very differently with a human implanted RFID chip. The good thing about this is that, these programs or software as they term it, is very much visible in the market as every developed RFID gadget is made along with it.

The software used by the RFID is greatly affected primarily by the compatibility issue. The compatibility factor between the device and the software plays a big role in impacting not only the gadget price but also the functionality and usability of it. For this reason, software developers and gadget manufacturers which plan to employ the RFID technology should work hand-in-hand to make sure that the issue of compatibility does not become a hindering factor after it has been made publicly available.

How RFID Technology Hits the Stock Market

The RFID has always been a big hit with companies wishing to acquire the technologically advanced tracking software offered. It is rather mind boggling to know though, that not much public research has been made on the subject behind the RFID stocks at the stock market. Very few investigations on the subject matter has ever been dealt with and reported to the public. However, news of big changes in the way RFID has been holding up with its stocks has been coming. Huge promise for RFID has yet to come, but the reports say that it will come for sure, and it will come with a huge impact.

Some of the few sources which have actually dealt an investigation on RFID stocks have promising news as well. Websites of the same sources have also been put up, but they are still rather fresh so more in depth information on investigations are still to be expected. Commentaries on the stocks of RFID have also been displayed on some website stock sources, and many of these are found on websites which only cover RFID stock investigative. Discussion boards have been put up in order to have a more in depth discussion between the public and other stock market investors and investigators.

Most topics discuss possible companies which are interested in implementing RFID technology with their line of products. Such topics speak of HP and Intermec, to name a few. RFID stocks have hit the stock market frenzy so much over the years that predicting future progression of its stocks has always excited a lot of companies.

How Stocks Research RFID Helps for Better Business

Radio-frequency identification is a new technology for automatic identification used by storing information in a small device and getting that information whenever needed. The proliferating use of the RFID technology has reached the meticulous system of business and enterprise. Today, most sectors of the society that use the RFID technology are commercially-motivated. Example of which are supermarkets, department stores, and groceries. This technology is slowly replacing the long use of barcodes to items. Instead of putting on barcodes to identify products, business owners now choose to put on RFID chips. This provides a more convenient tracking and inventory of their stocks, making an easier stocks research. In that case, most businesses choose to have stocks research RFID.

Stocks research RFID is an inventory method by using the RFID chips in tracking all the items in a certain store. In general, the word "stocks" is defined as a supply of goods that are kept on hand to be sold to customers. In finance, stocks may be defined as the outstanding capital of a certain corporation or company. During the days of barcode use, some stores have difficulties in tracking their stocks. Some even have inaccuracies on inventories. But everything has changed when RFID technology was used in stores.

In a study conducted in Wal-Mart, a popular convenient store worldwide, the use of stocks research RFID has helped the store a lot in terms of accuracy in inventories and savings. Since the use of stocks research RFID is simpler and easier, the management needs only fewer people to take care of the inventory process. The accurate inventory also led to their productive tracking of supplies and updated account of the stocks available. Due to the stocks research RFID, Wal-Mart has prevented out-of-stock issues.

Pointers When Looking for Your Supplier RFID

When it comes to looking for a supplier RFID, there are lots of considerations which must be thought of. In the first place, the tag involves a complex interplay of various networks so you cannot simply trust brands much more supplier RFID which have not properly proven their worth in the said industry. The first thing you needed to id you will purchase a tag is research your options. This will make you more exposed with the different aspects of the tag as well as the different industries where it is being widely used.

One good way of trimming down your supplier options is by getting in touch with people whom you know have been utilizing such tags. Even if you do not personally know them, feel free to drop them and email query and you will be surprised how eager they can in providing you with vital information. You can also begin to browse various IT forums online to try prying about discussions on such tags. You will definitely get loads of information through forums because this is a virtual place where people from a particular field hold their discussions.

When buying tags, make sure that you come face to face with the supplier of RFID. It is best that you be able to scrutinize the device and not just rely on mere pictures alone. Although you may find a potential supplier online, do not be too complacent to just get the tags shipped onto you. Make the extra effort of seeing it with your own two eyes.

Enhancing the Supply through Supply Chain RFID

The contemporary times gives the management of supply chain a thing that goes beyond counting up the boxes. This is the Radio Frequency Identification (RFID). Through the use of this application, the manufacturers can easily determine the details of their inventories aside from their numbers.

Formerly, Wal-Mart required all the suppliers to use bar codes in tracing their inventories. Bar codes basically trace the course of products from wharf to stock. But in today's world, this practice was revolutionized. The management of supply chain employs a new practice in tracing their inventories. Wal-Mart announces the use of supply chain RFID as a new method in tracing inventories.

RFID is a technology that is being enjoyed by many for almost 50 years now. This was first associated in tracing wildlife. The basic idea of this technology is very simple — to put a radio frequency transponder, which includes a microchip otherwise known as RFID tag, on things that are being traced; this will then emit an indication when it goes through a scanner. This technology was made affordable now so the supply chain can easily invest on this one.

RFID has more advantages as compared to the bar codes. Both are used for the same purpose — to trace. RFID however is more technological in nature that this can give more accurate and more information regarding the product specifications. Like for instance, this is capable in answering the questions like what is the product, where will it be delivered, and how will this be handled. Therefore, RFID is more capable in storing more information than the bar code.

Uses of Tags RFID

When the term RFID is defined, it will mean the general concept of the radio-frequency identification technology. Sometimes, people mistakenly identify this with the RFID device that is put on the product, person, or animal for the purpose of identifying or tracking them. But this device is not just called as the RFID. It is properly called as the tags RFID.

Tags RFID are just small devices, even smaller than the barcode sticker. Its size may range from a centimetre long and 0.3 cm wide. Some tags RFID are so small like the size of a pea. But this small device is so powerful that it helped the business sectors do their business better and make income of profits better. With the use of the tags RFID, products sold by different companies can have their own identities and therefore, inventories and assessment of them is easier and faster. Out-of-stock issues can also be prevented with these small devices.

In most cases, the tags RFID has only simple data like the name of the product and its identification number. But some advanced tags RFID may contain more data about the person, animal, or product. For example, those tags RFID used for persons like in their passports may contain their name, address, age, date of departure and arrival, destination, and their picture. Some may even include their different credit card numbers and other personal data.

The efficacy of the tags RFID also varies on its type. There are three major variation of this device: the active RFID, passive RFID, and the semi-passive RFID. The most effective in data storage and most reliable one is the active. Meanwhile, the commonly used is the passive RFID. However, the most significant can be the semi-passive which has both the benefits of the two other types of tags RFID.

Ubiquitous Computing Internet RFID: The Vision for Tomorrow

Businessmen and ordinary people alike have seen what RFID can do. They have seen its uses and the benefits that this technology brings. While it has its own share of criticisms and there are quite a few concerns voiced out by some people and by some sectors, the benefits of implementing such technology are just quite tempting.

More and more improvements and enhancements are being offered every day. And the products of RFID research have just shown the world how the possibilities of applications of such technology are just great. It would not be a surprise if one company comes up with something no one ever thought of. Technology developments just work that way. People saw this when the internet was introduced. And look at how much the internet has evolved since then.

And yet the trend of today is having products that enable people to do things wherever they are. A good example would be the cellular phone. At first it allowed people to talk through phones even when they were not at home. Then it was enabled to take pictures. And now it can even access the internet.

But what we are going to is not just being able to gain access to or communicating with people. The direction that technology is heading towards is connecting with objects through the use of RFID technology. The network of every person would even extend to the things that mean to him. It may be just simple things such as his shoes, clothes or maybe even his toothbrush. This will be the way of living of tomorrow. And people better be prepared because ubiquitous computing internet RFID would soon be knocking at their doorsteps.

The Essential RFID Ubiquitous Computing Code

The centre behind the development of the multi code tags found in the RFID microchips, intelligent card tags, and the popular bar codes is the Ubiquitous ID Centre. It provides the technology behind these tracker devices and software. The Ucode tags which Ubiquitous ID Centre provides make use of 128-bit of code which is picked up by the reading devices for Ucodes on tags. Reading and tracking devices come in different forms. What used to be big and bulky gadgets of the past, the reason why they were not as popular back then as they are today, would come in small devices the size of your palm, or sometimes even smaller. What is read from Ucode tags is the data which are specific for the current item the Ucode tag is attached to. The readers only need to use an infrared sensor to be able to read the contents of each Ucode. This same form of technology is also being used not just in code readers, but also in home appliances today like the television.

The smaller and more efficient both the Ucode tags and Ucode readers come these days, the better. With today's innovative technology, tags which used to make use of batteries are no more. These days, tags such as the latest RFID tags make use of small micro-processors which no more has the need to use batteries. The tags have been improving ever since, as well as getting much more affordable for companies such as supply lines to use with their products. RFID tags these days have gone from big, clunky battery powered tags, to small bar codes on a piece of sticker.

Wal-Mart And RFID Together Reach New Heights

It is the aim of every enterprise to be more efficient and to be more productive in order to gain a bigger income and to have growth in the company. As new technologies arrive businesses are willing to explore the possibilities of implementing such technology in order to improve their performance. With the development of the RFID technology, many businesses have proven what a great help it was in achieving more efficiency in the company. And so it shouldn't even be a surprise that many other enterprises have followed this example.

Wal-Mart is one of the early adapters to this technology. Through a study done in the company they have seen a great improvement in inventory management through the use of the RFID technology. There was a decrease of up to 30% in out-of-stocks. This affected their performance not only internally but also to their service to their customers. It was also easier now to monitor inventory. And not only that, the inaccuracies there were reduced too. Labour cost was also reduced due to the fact that it was possible to track inventory without human intervention.

In connection to its implementation of the RFID, Wal-Mart required its vendors and suppliers to send in their products with RFID labels on them. The vendors had to use RFID printers in order to put the RFID labels on the cases and pallets of the product that needed electronic product code. This move actually enforced the implementation of RFID in many companies. But with their compliance they were able to continue doing business with Wal-Mart. Wal-Mart, in turn, became more efficient in supply chain management.

Wal-Mart: Understanding RFID mandate

Founded in 1962, Wal-mart is an American corporation that manages a group of large companies worldwide. It was not until October 31, 1969 that Wal-Mart was officially incorporated. Since then, Wal-mart expanded to becoming United State's largest grocery retailer and the world's largest private employer. Wal-mart functions in UK (ASDA), Mexico (as Walmex), Japan (as Seiyu), Argentina, Canada, Brazil, Puerto Rico, China, South America, Germany, and South Korea.

It was Wal-mart's joint effort with the United States Department of Defence (USDD) that led to the formulation and publication of set standard for their vendors; obliging vendors to install RFID tags on all their shipments for to achieve an enhanced supply chain management (SCM). SCM is defined as the procedure of planning, carrying out, and directing operations of the supply chain. It normally covers all movement, storage, inventory, and completion of goods; from production stage to possibly consumption.

Apparently, due to Wal-Mart and USDD's enormous size, their formulated RFID mandates caused a significant impact on thousands of businesses or companies worldwide. Settled deadlines were often delayed since many of their vendors have had difficulties in carrying out RFID systems. Despite this scenario, RFID systems are expected to be adopted as well by even small companies globally.

The RFID mandates specifically required vendors to use RFID labels throughout their shipments. Vendors were basically obliged to use RFID printer/encoders in labelling their pallets and cases that needed for Wal-Mart's EPC (electronic product code) tags or smart labels. EPC is essentially a low-cost means of tracking and monitoring goods or products.

Through the years, Wal-Mart may have confronted various critics and issues. However, these were never considered as major concerns or obstacles to achieving further growth and success, particularly with its RFID systems adoption standards.

Malls like Wal-Mart Might Implement RFID Soon

With the RFID technology already creating groundbreaking innovative software for the industry, it won't be long before shopping in any mall becomes a trip similar to a trip to a friend's personal store. This is ultimately possible with the way RFID technology has been invading multiply fields of industry. It would not be long before even your own clothing is integrated with the software chip of RFID. Walking into a store such as Wal-Mart, which it is also expected that mall chains such as this will also join in the RFID craze in the future, we might expect sensors by the entrance to pick up the data stored in the RFID chip of your clothing, allowing the store employees to get to know you more with just a look at a monitor or projector. Imagine your past Wal-Mart purchases reflected on that screen, as well as your personal information, like your name. It would be easy for store employees to greet you by your name and they can even recommend store items which you might like, and chances are they might just be what you are actually looking for in the first place!

It might just be that a customer's every movement while inside the store perimeter is watched by the mall system. Upon purchasing your items from the mall, the checkout clerk might even recommend that you subscribe with them to get their store's credit card because they know, based on your store customer data, that you have been paying for your previous purchases in cash. That is truly something to look forward to, and is not that much impossible, with the way RFID is turning things.

Better Warehouse Efficiency with RFID

When it comes to warehouse systems, using radio frequency identification can do a lot to improve their work. When these companies use radio waves in order to track down certain objects, they also provide practically real time views of the status as well as the location of the product itself, thus making radio frequency identification the factor which greatly improves supply chains. Radio frequency identification is the best thing since sliced bread when it comes to getting instantaneous data which will be able to tell which particular products have actually been sold and how many more pieces or units are left on the shelves as well as in distribution centres and the warehouses. Why is this essential for keeping track of the progress of warehouse transactions? It provides control over information in a very good inventory way and makes smoother distribution channels, thus making management a lot more efficient and a lot less costly.

Take for example a classic warehouse scenario wherein a forklift goes to pick up a pallet that is loaded with a couple dozen cartons of the product which came from the warehouse. If one were to use the bar code technology, then each package that is found on the pallet will require to have its label to be scanned in a manual way in order to track what is being moved around. How time-wasteful is that? Using radio frequency identification technology will allow you to do the same tracking by making sure to install the radio frequency identification scanner in the doorway itself (a little bit like tollbooths on express lanes) and registers everything found in the load without having to lift the items one by one.

RFID: What is it and How Is It Used Nowadays?

During the course of World War II and the following years to come, devices that used sound and radio waves were made for different functions. One device in UK was invented at the course of World War II to identify whether an aircraft was ally or foe. In Russia, a device was made which has transmitter radio waves. This has audio information and was used for espionage by the Russians.

This is not an actual RFID, but it is a predecessor of what is now known as the RFID or Radio Frequency Identification. RFID is a device that relies on whose identification relies on automatic detection. How is this possible? It is possible by relying on the storage of some data and then remotely retrieving the said data using devices called RFID tags or "transponders".

There are 3 main types of RFID tags and they are the passive (not needing external power), active and semi-passive (both needing external power like a small battery). Passive RFID tags have no internal means of power supply. In contrast to this type of RFID tag, the active RFID has internal source of power. This source sustains power to the integrated circuits and transmits signal to the reader. The last type of RFID tag which is the semi-passive is a combination of both. It does have a power source which comes from a battery, which will power the microchip only. It will not, however, transmit signal.

Today, the use of RFID has gone beyond the military use. It is now being used in many fields like in medicine, consumer goods, public transportation, passports, product tracking and so many others. Indeed, RFID has come a long way from being just a simple identification or espionage device.

Wi-Fi RFID: Product of Technology Combination

Wi-Fi and RFID are closely related due to their functions and the manner that they are utilized. The RFID or radio-frequency identification technology is used for products, animals, or humans to provide them unique identification. That way, these tagged items will be recognized by a special RFID reader. Since the device used for this technology has data stored in it, reading devices can easily get the information through a simple swiping or lighting of the product. That is also a wireless technology. Now, the Wi-Fi or wireless fidelity also works just the same. As the name implies, it is also a wireless technology providing wireless internet or other network connections within areas capable to received Wi-Fi power. Due to the almost similar characteristic, RFID and Wi-Fi are now capable to become one — that being the Wi-Fi RFID.

The most common problem with the RFID is the limited area where it can be used. Mostly, RFID tags can only be recognized within few inches. But with the Wi-Fi RFID technology, RFID can now be used even in a far-reaching area. This will make the RFID mobile and move outside its limited environment. With the Wi-Fi RFID, tags can now be read by commercially available Wi-Fi APs.

This advanced technology was initiated by the Aeroscout Inc. This move changed the simple RFID into a special-purpose device for another application available in a Wi-Fi network. Through that, technology can now use the high-speed Wi-Fi with the cost effective RFID.

So far, the Wi-Fi RFID is still on its track for a more advanced use. Researchers are making new ways on how the small device can be helpful for the human kind in the near future.

Basics of the Wireless Mesh RFID eVectis

There is no doubt that the use of RFID or radio-frequency identification is very common now especially in the Western nations. Most consumers who often buy known brands might have encountered using the RFID tags or those devices put on an item for their unique identification. But the use of RFID is not only limited to commerce. Now, RFID has many future uses that are slowly realized. One of which is the use of this technology with the wireless mesh through the effort of eVectis Technologies, a company providing IT solutions for businesses. Thus, this entire setup is called the wireless mesh RFID eVectis.

As the use of RFID becomes wider and wider, people have realized that this technology can be utilized as a wireless communication technology. It is known that the RFID tags are capable of sending messages or data even in a far distance. That is, of course, depends on the type of RFID tag used. Nevertheless, the use of RFID tags in wireless communication has now become a reality. In fact, wireless mesh is now using RFID technology. And the most known company providing this is the eVectis Technologies.

The wireless mesh is a network of mesh created using a wireless connection device that is installed to each of the network's connected computer. With this type of networking, every unit connected to the network is a provider and at the same time a receiver of data. The unique about this setup is that every computer here is connected to one another. This is needed in order to use the RFID technology more effectively.

This kind of network setup is usually done through the help of eVectis. This company is actually the most trusted in terms of wireless mesh RFID eVectis.

Wireless Mesh RFID Evectis.Com: Taking the RFID Technology One Step Further

The wireless mesh RFID of evictis.com marries two technologies: RFID delivery of data and VoIP capability using Wi-Fi solutions. This is a step forward in the RFID technology systems because of its effective interface with a computer system in real time. Wire meshing of RFID signals uses a different algorithm so that radio frequencies can be transmitted to mobile devices such as laptop, a mini base receiver, or mobile data reader. This variant of RFID can provide superior data transmission in real time but at the same time has a capability to transmit other signals such as audio and video broadcast comparable to most broadband technologies.

Wireless mesh RFID is ideal for surveillance especially on mobile platforms. This would ensure direct video signal transmission to a configured mobile device such as a laptop computer. So users can set continued video surveillance even while in transit. This adds new dimensions to the security capabilities of the RFID systems. It is also ideal for interconnecting users on a wireless mesh networks. Its practical uses could be in the form transmitting general advisories for those who are part of the network. This will significantly ease communication and coordination of activities.

Wireless mesh RFID is a significant step in both wireless technology and radio frequency modulated data transmission. The technology could greatly easy management and processing of data. Its effective interface to mobile devices together with its built-in ability to send and receive multi-media streams magnifies its importance on several security issues hounding an enterprise.

Best RFID Solutions from Zebra

When it comes to your radio frequency identification needs, there is a company that you can trust one hundred percent, and that company is Zebra. This is because such a company is able to understand and provide solutions of companies that are concerned with or have a need for implementation of a radio frequency identification solution. Even if it is just simple navigation of the different retailer requirements or government mandates, Zebra goes all out to help you with the basic tenets of radio frequency identification solutions, even going as far as to specify and order the correct smart labels as well as supplies. Clearly, Zebra ought to be your RFID partner of choice.

For some time now, Zebra has proven its worth as a leading provider of rugged as well as reliable printing solutions. They also have good on-demand printing of Zebra is also the most recognized brand heralded in the industry of automatic identification because they have the most complete and up to date product line. More than ninety percent of the companies that are found in the Fortune 500 companies list use Zebra solutions as well, proving that only be best companies and organizations also use the best media solutions that come from the fastest developing companies for radio frequency identification as well.